Hands-on Guide to Apache Spark 3

Build Scalable Computing Engines for Batch and Stream Data Processing

Alfonso Antolínez García

Apress®

Hands-on Guide to Apache Spark 3: Build Scalable Computing Engines for Batch and Stream Data Processing

Alfonso Antolínez García
Madrid, Spain

ISBN-13 (pbk): 978-1-4842-9379-9
https://doi.org/10.1007/978-1-4842-9380-5

ISBN-13 (electronic): 978-1-4842-9380-5

Managing Director, Apress Media LLC: Welmoed Spahr
Acquisitions Editor: Celestin Suresh John
Development Editor: Laura Berendson
Coordinating Editor: Mark Powers

Cover designed by eStudioCalamar

Cover image by Lucas Santos on Unsplash (www.unsplash.com)

Distributed to the book trade worldwide by Apress Media, LLC, 1 New York Plaza, New York, NY 10004, U.S.A. Phone 1-800-SPRINGER, fax (201) 348-4505, e-mail orders-ny@springer-sbm.com, or visit www.springeronline.com. Apress Media, LLC is a California LLC and the sole member (owner) is Springer Science + Business Media Finance Inc (SSBM Finance Inc). SSBM Finance Inc is a **Delaware** corporation.

For information on translations, please e-mail booktranslations@springernature.com; for reprint, paperback, or audio rights, please e-mail bookpermissions@springernature.com.

Apress titles may be purchased in bulk for academic, corporate, or promotional use. eBook versions and licenses are also available for most titles. For more information, reference our Print and eBook Bulk Sales web page at http://www.apress.com/bulk-sales.

Any source code or other supplementary material referenced by the author in this book is available to readers on GitHub (https://github.com/Apress). For more detailed information, please visit http://www.apress.com/source-code.

Printed on acid-free paper

To my beloved family

Table of Contents

About the Author

 **Alfonso Antolínez García** is a senior IT manager with a long professional career serving in several multinational companies such as Bertelsmann SE, Lafarge, and TUI AG. He has been working in the media industry, the building materials industry, and the leisure industry. Alfonso also works as a university professor, teaching artificial intelligence, machine learning, and data science. In his spare time, he writes research papers on artificial intelligence, mathematics, physics, and the applications of information theory to other sciences.

About the Technical Reviewer

Akshay R. Kulkarni is an AI and machine learning evangelist and a thought leader. He has consulted several Fortune 500 and global enterprises to drive AI- and data science–led strategic transformations. He is a Google Developer Expert, author, and regular speaker at major AI and data science conferences (including Strata, O'Reilly AI Conf, and GIDS). He is a visiting faculty member for some of the top graduate institutes in India. In 2019, he has been also featured as one of the top 40 under-40 data scientists in India. In his spare time, he enjoys reading, writing, coding, and building next-gen AI products.

PART I

Apache Spark Batch Data Processing

CHAPTER 1

Introduction to Apache Spark for Large-Scale Data Analytics

Apache Spark started as a research project at the UC Berkeley AMPLab in 2009. It became open source in 2010 and was transferred to the Apache Software Foundation in 2013 and boasts the largest open source big data community.

From its genesis, Spark was designed with a significant change in mind, to store intermediate data computations in Random Access Memory (RAM), taking advantage of the coming-down RAM prices that occurred in the 2010s, in comparison with Hadoop that keeps information in slower disks.

In this chapter, I will provide an introduction to Spark, explaining how it works, the Spark Unified Analytics Engine, and the Apache Spark ecosystem. Lastly, I will describe the differences between batch and streaming data.

1.1 What Is Apache Spark?

Apache Spark is a unified engine for large-scale data analytics. It provides high-level application programming interfaces (APIs) for Java, Scala, Python, and R programming languages and supports SQL, streaming data, machine learning (ML), and graph processing. Spark is a multi-language engine for executing data engineering, data science, and machine learning on single-node machines or clusters of computers, either on-premise or in the cloud.

© Alfonso Antolínez García 2023
A. Antolínez García, *Hands-on Guide to Apache Spark 3*, https://doi.org/10.1007/978-1-4842-9380-5_1

Spark provides in-memory computing for intermediate computations, meaning data is kept in memory instead of writing it to slow disks, making it faster than Hadoop MapReduce, for example. It includes a set of high-level tools and modules such as follows: Spark SQL is for structured data processing and access to external data sources like Hive; MLlib is the library for machine learning; GraphX is the Spark component for graphs and graph-parallel computation; Structured Streaming is the Spark SQL stream processing engine; Pandas API on Spark enables Pandas users to work with large datasets by leveraging Spark; SparkR provides a lightweight interface to utilize Apache Spark from the R language; and finally PySpark provides a similar front end to run Python programs over Spark.

There are five key benefits that make Apache Spark unique and bring it to the spotlight:

- Simpler to use and operate

- Fast

- Scalable

- Ease of use

- Fault tolerance at scale

Let's have a look at each of them.

Simpler to Use and Operate

Spark's capabilities are accessed via a common and rich API, which makes it possible to interact with a unificd general-purpose distributed data processing engine via different programming languages and cope with data at scale. Additionally, the broad documentation available makes the development of Spark applications straightforward.

The Hadoop MapReduce processing technique and distributed computing model inspired the creation of Apache Spark. This model is conceptually simple: divide a huge problem into smaller subproblems, distribute each piece of the problem among as many individual solvers as possible, collect the individual solutions to the partial problems, and assemble them in a final result.

Fast

On November 5, 2014, Databricks officially announced they have won the Daytona GraySort contest.[1] In this competition, the Databricks team used a Spark cluster of 206 EC2 nodes to sort 100 TB of data (1 trillion records) in 23 minutes. The previous world record of 72 minutes using a Hadoop MapReduce cluster of 2100 nodes was set by Yahoo. Summarizing, Spark sorted the same data three times faster with ten times fewer machines. Impressive, right?

But wait a bit. The same post also says, *"All the sorting took place on disk (HDFS), without using Spark's in-memory cache."* So was it not all about Spark's in-memory capabilities? Apache Spark is recognized for its in-memory performance. However, assuming Spark's outstanding results are due to this feature is one of the most common misconceptions about Spark's design. From its genesis, Spark was conceived to achieve a superior performance both in memory and on disk. Therefore, Spark operators perform regular operations on disk when data does not fit in memory.

Scalable

Apache Spark is an open source framework intended to provide parallelized data processing at scale. At the same time, Spark high-level functions can be used to carry out different data processing tasks on datasets of diverse sizes and schemas. This is accomplished by distributing workloads from several servers to thousands of machines, running on a cluster of computers and orchestrated by a cluster manager like Mesos or Hadoop YARN. Therefore, hardware resources can increase linearly with every new computer added. It is worth clarifying that hardware addition to the cluster does not necessarily represent a linear increase in computing performance and hence linear reduction in processing time because internal cluster management, data transfer, network traffic, and so on also consume resources, subtracting them from the effective Spark computing capabilities. Despite the fact that running in cluster mode leverages Spark's full distributed capacity, it can also be run locally on a single computer, called local mode.

---

[1] www.databricks.com/blog/2014/11/05/spark-officially-sets-a-new-record-in-large-scale-sorting.html

If you have searched for information about Spark before, you probably have read something like "Spark runs on commodity hardware." It is important to understand the term "commodity hardware." In the context of big data, commodity hardware does not denote low quality, but rather equipment based on market standards, which is general-purpose, widely available, and hence affordable as opposed to purpose-built computers.

Ease of Use

Spark makes the life of data engineers and data scientists operating on large datasets easier. Spark provides a single unified engine and API for diverse use cases such as streaming, batch, or interactive data processing. These tools allow it to easily cope with diverse scenarios like ETL processes, machine learning, or graphs and graph-parallel computation. Spark also provides about a hundred operators for data transformation and the notion of dataframes for manipulating semi-structured data.

Fault Tolerance at Scale

At scale many things can go wrong. In the big data context, fault refers to failure, that is to say, Apache Spark's fault tolerance represents its capacity to operate and to recover after a failure occurs. In large-scale clustered environments, the occurrence of any kind of failure is certain at any time; thus, Spark is designed assuming malfunctions are going to appear sooner or later.

Spark is a distributed computing framework with built-in fault tolerance that takes advantage of a simple data abstraction named a RDD (Resilient Distributed Dataset) that conceals data partitioning and distributed computation from the user. RDDs are immutable collections of objects and are the building blocks of the Apache Spark data structure. They are logically divided into portions, so they can be processed in parallel, across multiple nodes of the cluster.

The acronym RDD denotes the essence of these objects:

- *Resilient (fault-tolerant)*: The RDD lineage or Directed Acyclic Graph (DAG) permits the recomputing of lost partitions due to node failures from which they are capable of recovering automatically.

- *Distributed*: RDDs are processes in several nodes in parallel.

- *Dataset*: It's the set of data to be processed. Datasets can be the result of parallelizing an existing collection of data; loading data from an external source such as a database, Hive tables, or CSV, text, or JSON files: and creating a RDD from another RDD.

Using this simple concept, Spark is able to handle a wide range of data processing workloads that previously needed independent tools.

Spark provides two types of fault tolerance: RDD fault tolerance and streaming write-ahead logs. Spark uses its RDD abstraction to handle failures of worker nodes in the cluster; however, to control failures in the driver process, Spark 1.2 introduced write-ahead logs, to save received data to a fault-tolerant storage, such as HDFS, S3, or a similar safeguarding tool.

Fault tolerance is also achieved thanks to the introduction of the so-called DAG, or Directed Acyclic Graph, concept. Formally, a DAG is defined as a set of vertices and edges. In Spark, a DAG is used for the visual representation of RDDs and the operations being performed on them. The RDDs are represented by vertices, while the operations are represented by edges. Every edge is directed from an earlier state to a later state. This task tracking contributes to making fault tolerance possible. It is also used to schedule tasks and for the coordination of the cluster worker nodes.

1.2 Spark Unified Analytics Engine

The idea of platform integration is not new in the world of software. Consider, for example, the notion of Customer Relationship Management (CRM) or Enterprise Resource Planning (ERP). The idea of unification is rooted in Spark's design from inception. On October 28, 2016, the Association for Computing Machinery (ACM) published the article titled "Apache Spark: a unified engine for big data processing." In this article, authors assert that due to the nature of big data datasets, a standard pipeline must combine MapReduce, SQL-like queries, and iterative machine learning capabilities. The same document states Apache Spark combines batch processing capabilities, graph analysis, and data streaming, integrating a single SQL query engine formerly split up into different specialized systems such as Apache Impala, Drill, Storm, Dremel, Giraph, and others.

Spark's simplicity resides in its unified API, which makes the development of applications easier. In contrast to previous systems that required saving intermediate data to a permanent storage to transfer it later on to other engines, Spark incorporates many functionalities in the same engine and can execute different modules to the same data and very often in memory. Finally, Spark has facilitated the development of new applications, such as scaling iterative algorithms, integrating graph querying and algorithms in the Spark Graph component.

The value added by the integration of several functionalities into a single system can be seen, for instance, in modern smartphones. For example, nowadays, taxi drivers have replaced several devices (GPS navigator, radio, music cassettes, etc.) with a single smartphone. In unifying the functions of these devices, smartphones have eventually enabled new functionalities and service modalities that would not have been possible with any of the devices operating independently.

1.3 How Apache Spark Works

We have already mentioned Spark scales by distributing computing workload across a large cluster of computers, incorporating fault tolerance and parallel computing. We have also pointed out it uses a unified engine and API to manage workloads and to interact with applications written in different programming languages.

In this section we are going to explain the basic principles Apache Spark uses to perform big data analysis under the hood. We are going to walk you through the Spark Application Model, Spark Execution Model, and Spark Cluster Model.

Spark Application Model

In MapReduce, the highest-level unit of computation is the job; in Spark, the highest-level unit of computation is the application. In a job we can load data, apply a map function to it, shuffle it, apply a reduce function to it, and finally save the information to a fault-tolerant storage device. In Spark, applications are self-contained entities that execute the user's code and return the results of the computation. As mentioned before, Spark can run applications using coordinated resources of multiple computers. Spark applications can carry out a single batch job, execute an iterative session composed of several jobs, or act as a long-lived streaming server processing unbounded streams of data. In Spark, a job is launched every time an application invokes an action.

Unlike other technologies like MapReduce, which starts a new process for each task, Spark applications are executed as independent processes under the coordination of the SparkSession object running in the driver program. Spark applications using iterative algorithms benefit from dataset caching capabilities among other operations. This is feasible because those algorithms conduct repetitive operations on data. Finally, Spark applications can maintain steadily running processes on their behalf in cluster nodes even when no job is being executed, and multiple applications can run on top of the same executor. The former two characteristics combined leverage Spark rapid startup time and in-memory computing.

Spark Execution Model

The Spark Execution Model contains vital concepts such as the driver program, executors, jobs, tasks, and stages. Understanding of these concepts is of paramount importance for fast and efficient Spark application development. Inside Spark, tasks are the smallest execution unit and are executed inside an executor. A task executes a limited number of instructions. For example, loading a file, filtering, or applying a map() function to the data could be considered a task. Stages are collections of tasks running the same code, each of them in different chunks of a dataset. For example, the use of functions such as reduceByKey(), Join(), etc., which require a shuffle or reading a dataset, will trigger in Spark the creation of a stage. Jobs, on the other hand, comprise several stages.

Next, due to their relevance, we are going to study the concepts of the driver program and executors together with the Spark Cluster Model.

Spark Cluster Model

Apache Spark running in cluster mode has a master/worker hierarchical architecture depicted in Figure 1-1 where the driver program plays the role of master node. The Spark Driver is the central coordinator of the worker nodes (slave nodes), and it is responsible for delivering the results back to the client. Workers are machine nodes that run executors. They can host one or multiple workers, they can execute only one JVM (Java Virtual Machine) per worker, and each worker can generate one or more executors as shown in Figure 1-2.

The Spark Driver generates the SparkContext and establishes the communication with the Spark Execution environment and with the cluster manager, which provides resources for the applications. The Spark Framework can adopt several cluster managers: Spark's Standalone Cluster Manager, Apache Mesos, Hadoop YARN, or Kubernetes. The driver connects to the different nodes of the cluster and starts processes called executors, which provide computing resources and in-memory storage for RDDs. After resources are available, it sends the applications' code (JAR or Python files) to the executors acquired. Finally, the SparkContext sends tasks to the executors to run the code already placed in the workers, and these tasks are launched in separate processor threads, one per worker node core. The SparkContext is also used to create RDDs.

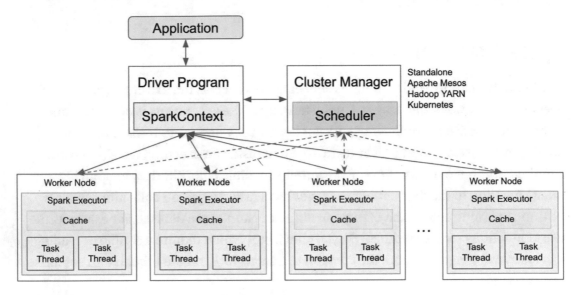

Figure 1-1. *Apache Spark cluster mode overview*

In order to provide applications with logical fault tolerance at both sides of the cluster, each driver schedules its own tasks and each task, running in every executor, executes its own JVM (Java Virtual Machine) processes, also called executor processes. By default executors run in static allocation, meaning they keep executing for the entire lifetime of a Spark application, unless dynamic allocation is enabled. The driver, to keep track of executors' health and status, receives regular heartbeats and partial execution metrics for the ongoing tasks (Figure 1-3). Heartbeats are periodic messages (every 10 s by default) from the executors to the driver.

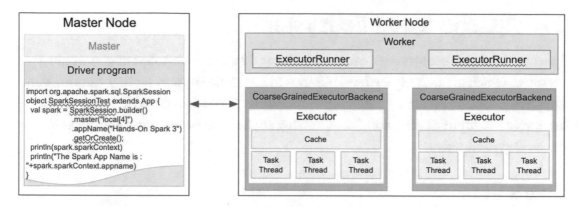

Figure 1-2. *Spark communication architecture with worker nodes and executors*

This Execution Model also has some downsides. Data cannot be exchanged between Spark applications (instances of the SparkContext) via the in-memory computation model, without first saving the data to an external storage device.

As mentioned before, Spark can be run with a wide variety of cluster managers. That is possible because Spark is a cluster-agnostic platform. This means that as long as a cluster manager is able to obtain executor processes and to provide communication among the architectural components, it is suitable for the purpose of executing Spark. That is why communication between the driver program and worker nodes must be available at all times, because the former must acquire incoming connections from the executors for as long as applications are executing on them.

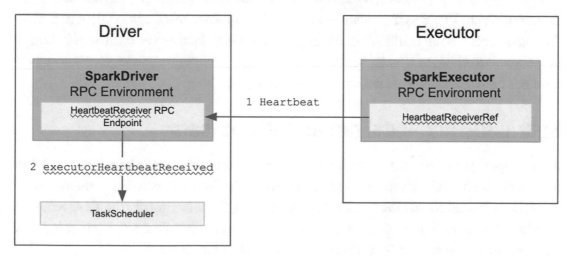

Figure 1-3. *Spark's heartbeat communication between executors and the driver*

1.4 Apache Spark Ecosystem

The Apache Spark ecosystem is composed of a unified and fault-tolerant core engine, on top of which are four higher-level libraries that include support for SQL queries, data streaming, machine learning, and graph processing. Those individual libraries can be assembled in sophisticated workflows, making application development easier and improving productivity.

Spark Core

Spark Core is the bedrock on top of which in-memory computing, fault tolerance, and parallel computing are developed. The Core also provides data abstraction via RDDs and together with the cluster manager data arrangement over the different nodes of the cluster. The high-level libraries (Spark SQL, Streaming, MLlib for machine learning, and GraphX for graph data processing) are also running over the Core.

Spark APIs

Spark incorporates a series of application programming interfaces (APIs) for different programming languages (SQL, Scala, Java, Python, and R), paving the way for the adoption of Spark by a great variety of professionals with different development, data science, and data engineering backgrounds. For example, Spark SQL permits the interaction with RDDs as if we were submitting SQL queries to a traditional relational database. This feature has facilitated many transactional database administrators and developers to embrace Apache Spark.

Let's now review each of the four libraries in detail.

Spark SQL and DataFrames and Datasets

Apache Spark provides a data programming abstraction called DataFrames integrated into the Spark SQL module. If you have experience working with Python and/or R dataframes, Spark DataFrames could look familiar to you; however, the latter are distributable across multiple cluster workers, hence not constrained to the capacity of a single computer. Spark was designed to tackle very large datasets in the most efficient way.

A DataFrame looks like a relational database table or Excel spreadsheet, with columns of different data types, headers containing the names of the columns, and data stored as rows as shown in Table 1-1.

Table 1-1. *Representation of a DataFrame as a Relational Table or Excel Spreadsheet*

firstName	lastName	profession	birthPlace
Antonio	Dominguez Bandera	Actor	Málaga
Rafael	Nadal Parera	Tennis Player	Mallorca
Amancio	Ortega Gaona	Businessman	Busdongo de Arbas
Pablo	Ruiz Picasso	Painter	Málaga
Blas	de Lezo	Admiral	Pasajes
Miguel	Serveto y Conesa	Scientist/Theologist	Villanueva de Sigena

On the other hand, Figure 1-4 depicts an example of a DataFrame.

```
+---------+------------------+--------------------+--------------------+
|firstName|          lastName|          profession|           bornPlace|
+---------+------------------+--------------------+--------------------+
|  Antonio|Dominguez Bandera|                Actor|              Malaga|
|   Rafael|     Nadal Parera|        Tennis Player|            Mallorca|
|  Amancio|     Ortega Gaona|          Businessman|   Busdongo de Arbas|
|    Pablo|     Ruiz Picasso|              Painter|              Malaga|
|     Blas|          de Lezo|              Admiral|             Pasajes|
|   Miguel| Serveto y Conesa|Scientist/Theologist|Villanueva de Sigena|
+---------+------------------+--------------------+--------------------+
```

Figure 1-4. *Example of a DataFrame*

A Spark DataFrame can also be defined as an integrated data structure optimized for distributed big data processing. A Spark DataFrame is also a RDD extension with an easy-to-use API for simplifying writing code. For the purposes of distributed data processing, the information inside a Spark DataFrame is structured around schemas. Spark schemas contain the names of the columns, the data type of a column, and its nullable properties. When the nullable property is set to true, that column accepts null values.

SQL has been traditionally the language of choice for many business analysts, data scientists, and advanced users to leverage data. Spark SQL allows these users to query structured datasets as they would have done if they were in front of their traditional data source, hence facilitating adoption.

On the other hand, in Spark a dataset is an immutable and a strongly typed data structure. Datasets, as DataFrames, are mapped to a data schema and incorporate type safety and an object-oriented interface. The Dataset API converts between JVM objects and tabular data representation taking advantage of the encoder concept. This tabular representation is internally stored in a binary format called Spark Tungsten, which improves operations in serialized data and improves in-memory performance.

Datasets incorporate compile-time safety, allowing user-developed code to be error-tested before the application is executed. There are several differences between datasets and dataframes. The most important one could be datasets are only available to the Java and Scala APIs. Python or R applications cannot use datasets.

Spark Streaming

Spark Structured Streaming is a high-level library on top of the core Spark SQL engine. Structured Streaming enables Spark's fault-tolerant and real-time processing of unbounded data streams without users having to think about how the streaming takes place. Spark Structured Streaming provides fault-tolerant, fast, end-to-end, exactly-once, at-scale stream processing. Spark Streaming permits express streaming computation in the same fashion as static data is computed via batch processing. This is achieved by executing the streaming process incrementally and continuously and updating the outputs as the incoming data is ingested.

With Spark 2.3, a new low-latency processing mode called continuous processing was introduced, achieving end-to-end latencies of as low as 1 ms, ensuring at-least-once[2] message delivery. The at-least-once concept is depicted in Figure 1-5. By default, Structured Streaming internally processes the information as micro-batches, meaning data is processed as a series of tiny batch jobs.

Figure 1-5. *Depiction of the at-least-once message delivery semantic*

---

[2] With the at-least-once message delivery semantic, a message can be delivered more than once; however, no message can be lost.

Spark Structured Streaming also uses the same concepts of datasets and DataFrames to represent streaming aggregations, event-time windows, stream-to-batch joins, etc. using different programming language APIs (Scala, Java, Python, and R). It means the same queries can be used without changing the dataset/DataFrame operations, therefore choosing the operational mode that best fits our application requirements without modifying the code.

Spark's machine learning (ML) library is commonly known as MLlib, though it is not its official name. MLlib's goal is to provide big data out-of-the-box, easy-to-use machine learning capabilities. At a high level, it provides capabilities such as follows:

- Machine learning algorithms like classification, clustering, regression, collaborative filtering, decision trees, random forests, and gradient-boosted trees among others

- Featurization:

  - Term Frequency-Inverse Document Frequency (TF-IDF) statistical and feature vectorization method for natural language processing and information retrieval.

  - *Word2vec*: It takes text corpus as input and produces the word vectors as output.

  - *StandardScaler*: It is a very common tool for pre-processing steps and feature standardization.

  - Principal component analysis, which is an orthogonal transformation to convert possibly correlated variables.

  - Etc.

- ML Pipelines, to create and tune machine learning pipelines

- Predictive Model Markup Language (PMML), to export models to PMML

- Basic Statistics, including summary statistics, correlation between series, stratified sampling, etc.

As of Spark 2.0, the primary Spark Machine Learning API is the DataFrame-based API in the spark.ml package, switching from the traditional RDD-based APIs in the spark.mllib package.

Spark GraphX

GraphX is a new high-level Spark library for graphs and graph-parallel computation designed to solve graph problems. GraphX extends the Spark RDD capabilities by introducing this new graph abstraction to support graph computation and includes a collection of graph algorithms and builders to optimize graph analytics.

The Apache Spark ecosystem described in this section is portrayed in Figure 1-6.

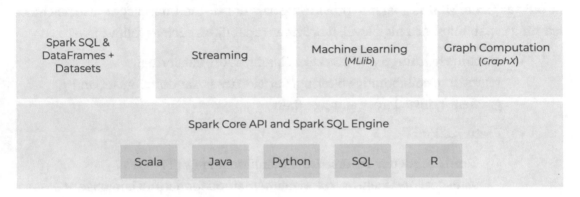

Figure 1-6. *The Apache Spark ecosystem*

In Figure 1-7 we can see the Apache Spark ecosystem of connectors.

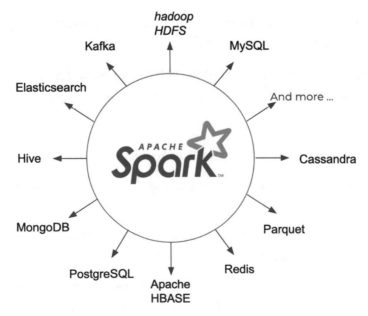

Figure 1-7. *Apache Spark ecosystem of connectors*

1.5 Batch vs. Streaming Data

Nowadays, the world generates boundless amounts of data, and it continues to augment at an astonishing rate. It is expected the volume of information created, captured, copied, and consumed worldwide from 2010 to 2025 will exceed 180 ZB.[3] If this figure does not say much to you, imagine your personal computer or laptop has a hard disk of 1 TB (which could be considered a standard in modern times). It would be necessary for you to have 163,709,046,319.13 disks to store such amount of data.[4]

Presently, data is rarely static. Remember the famous three Vs of big data:

- Volume

  The unprecedented explosion of data production means that storage is no longer the real challenge, but to generate actionable insights from within gigantic datasets.

- Velocity

  Data is generated at an ever-accelerating pace, posing the challenge for data scientists to find techniques to collect, process, and make use of information as it comes in.

- Variety

  Big data is disheveled, sources of information heterogeneous, and data formats diverse. While structured data is neatly arranged within tables, unstructured data is information in a wide variety of forms without following predefined data models, making it difficult to store in conventional databases. The vast majority of new data being generated today is unstructured, and it can be human-generated or machine-generated. Unstructured data is more difficult to deal with and extract value from. Examples of unstructured data include medical images, video, audio files, sensors, social media posts, and more.

---

[3] www.statista.com/statistics/871513/worldwide-data-created/

[4] 1 zettabyte = 10^{21} bytes.

For businesses, data processing is critical to accelerate data insights obtaining deep understanding of particular issues by analyzing information. This deep understanding assists organizations in developing business acumen and turning information into actionable insights. Therefore, it is relevant enough to trigger actions leading us to improve operational efficiency, gain competitive advantage, leverage revenue, and increase profits. Consequently, in the face of today's and tomorrow's business challenges, analyzing data is crucial to discover actionable insights and stay afloat and profitable.

It is worth mentioning there are significant differences between data insights, data analytics, and just data, though many times they are used interchangeably. Data can be defined as a collection of facts, while data analytics is about arranging and scrutinizing the data. Data insights are about discovering patterns in data. There is also a hierarchical relationship between these three concepts. First, information must be collected and organized, only after it can be analyzed and finally data insights can be extracted. This hierarchy can be graphically seen in Figure 1-8.

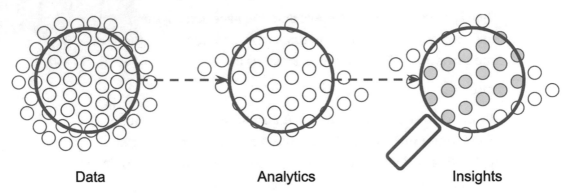

Data   Analytics   Insights

Figure 1-8. *Hierarchical relationship between data, data analytics, and data insight extraction*

When it comes to data processing, there are many different methodologies, though stream and batch processing are the two most common ones. In this section, we will explain the differences between these two data processing techniques. So let's define batch and stream processing before diving into the details.

What Is Batch Data Processing?

Batch data processing can be defined as a computing method of executing high-volume data transactions in repetitive batches in which the data collection is separate from the processing. In general, batch jobs do not require user interaction once the process is initiated. Batch processing is particularly suitable for end-of-cycle treatment of information, such as warehouse stock update at the end of the day, bank reconciliation, or monthly payroll calculation, to mention some of them.

Batch processing has become a common part of the corporate back office processes, because it provides a significant number of advantages, such as efficiency and data quality due to the lack of human intervention. On the other hand, batch jobs have some cons. The more obvious one could be they are complex and critical because they are part of the backbone of the organizations. As a result, developing sophisticated batch jobs can be expensive up front in terms of time and resources, but in the long run, they pay the investment off.

Another disadvantage of batch data processing is that due to its large scale and criticality, in case of a malfunction, significant production shutdowns are likely to occur. Batch processes are monolithic in nature; thus, in case of rise in data volumes or peaks of demand, they cannot be easily adapted.

What Is Stream Data Processing?

Stream data processing could be characterized as the process of collecting, manipulating, analyzing, and delivering up-to-date information and keeping the state of data updated while it is still in motion. It could also be defined as a low-latency way of collecting and processing information while it is still in transit. With stream processing, the data is processed in real time; thus, there is no delay between data collection and processing and providing instant response.

Stream processing is particularly suitable when data comes in as a continuous flow while changing over time, with high velocity, and real-time analytics is needed or response to an event as soon as it occurs is mandatory. Stream data processing leverages active intelligence, owning in-the-moment consciousness about important business events and enabling triggering instantaneous actions. Analytics is performed instantly on the data or within a fraction of a second; thus, it is perceived by the user as a real-time update. Some examples where stream data processing is the best option are credit card fraud detection, real-time system security monitoring, or the use of Internet-of-Things (IoT) sensors. IoT devices permit monitoring anomalies in machinery and provide control with a heads-up as soon as anomalies or outliers[5] are detected. Social media and customer sentiment analysis are other trendy fields of stream data processing application.

One of the main disadvantages of stream data processing is implementing it at scale. In real life data streaming is far away from being perfect, and often data does not flow regularly or smoothly. Imagine a situation in which data flow is disrupted and some data is missing or broken down. Then, after normal service is restored, that missing or broken-down information suddenly arrives at the platform, flooding the processing system. To be able to cope with situations like this, streaming architectures require spare capacity of computing, communications, and storage.

Difference Between Stream Processing and Batch Processing

Summarizing, we could say that stream processing involves the treatment and analysis of data in motion in real or almost-real time, while batch processing entails handling and analyzing static information at time intervals.

In batch jobs, you manipulate information produced in the past and consolidated in a permanent storage device. It is also what is commonly known as information at rest.

In contrast, stream processing is a low-latency solution, demanding the analysis of streams of information while it is still in motion. Incoming data requires to be processed in flight, in real or almost-real time, rather than saved in a permanent storage. Given that data is consumed as it is generated, it provides an up-to-the-minute snapshot of the information, enabling a proactive response to events. Another difference between batch and stream processing is that in stream processing only the information considered relevant for the process being analyzed is stored from the very beginning. On the other hand, data considered of no immediate interest can be stored in low-cost devices for ulterior analysis with data mining algorithms, machine learning models, etc.

A graphical representation of batch vs. stream data processing is shown in Figure 1-9.

---

[5] Parameters out of defined thresholds.

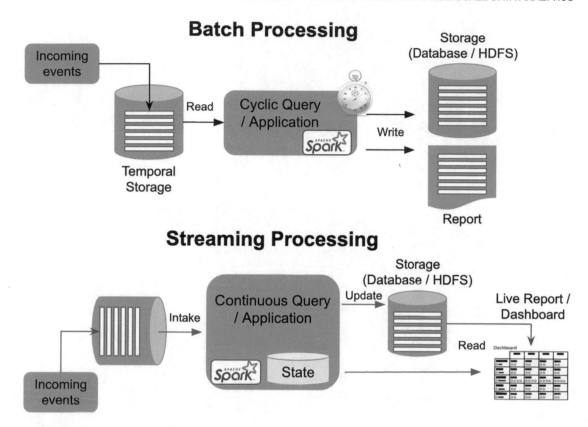

Figure 1-9. *Batch vs. streaming processing representation*

1.6 Summary

In this chapter we briefly looked at the Apache Spark architecture, implementation, and ecosystem of applications. We also covered the two different types of data processing Spark can deal with, batch and streaming, and the main differences between them. In the next chapter, we are going to go through the Spark setup process, the Spark application concept, and the two different types of Apache Spark RDD operations: transformations and actions.

CHAPTER 2

Getting Started with Apache Spark

Now that you have an understanding of what Spark is and how it works, we can get you set up to start using it. In this chapter, I'll provide download and installation instructions and cover Spark command-line utilities in detail. I'll also review Spark application concepts, as well as transformations, actions, immutability, and lazy evaluation.

2.1 Downloading and Installing Apache Spark

The first step you have to take to have your Spark installation up and running is to go to the Spark download page and choose the Spark release 3.3.0. Then, select the package type "Pre-built for Apache Hadoop 3.3 and later" from the drop-down menu in step 2, and click the "Download Spark" link in step 3 (Figure 2-1).

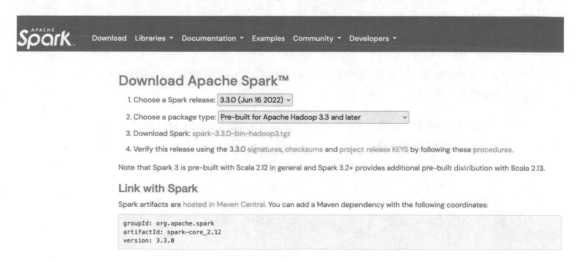

Figure 2-1. *The Apache Spark download page*

© Alfonso Antolínez García 2023
A. Antolínez García, *Hands-on Guide to Apache Spark 3*, https://doi.org/10.1007/978-1-4842-9380-5_2

This will download the file `spark-3.3.0-bin-hadoop3.tgz` or another similar name in your case, which is a compressed file that contains all the binaries you will need to execute Spark in local mode on your local computer or laptop.

What is great about setting Apache Spark up in local mode is that you don't need much work to do. We basically need to install Java and set some environment variables. Let's see how to do it in several environments.

Installation of Apache Spark on Linux

The following steps will install Apache Spark on a Linux system. It can be Fedora, Ubuntu, or another distribution.

Step 1: Verifying the Java Installation

Java installation is mandatory in installing Spark. Type the following command in a terminal window to verify Java is available and its version:

```
$ java -version
```

If Java is already installed on your system, you get to see a message similar to the following:

```
$ java -version
java version "18.0.2" 2022-07-19
Java(TM) SE Runtime Environment (build 18.0.2+9-61)
Java HotSpot(TM) 64-Bit Server VM (build 18.0.2+9-61, mixed mode, sharing)
```

Your Java version may be different. Java 18 is the Java version in this case.

If you don't have Java installed

1. Open a browser window, and navigate to the Java download page as seen in Figure 2-2.

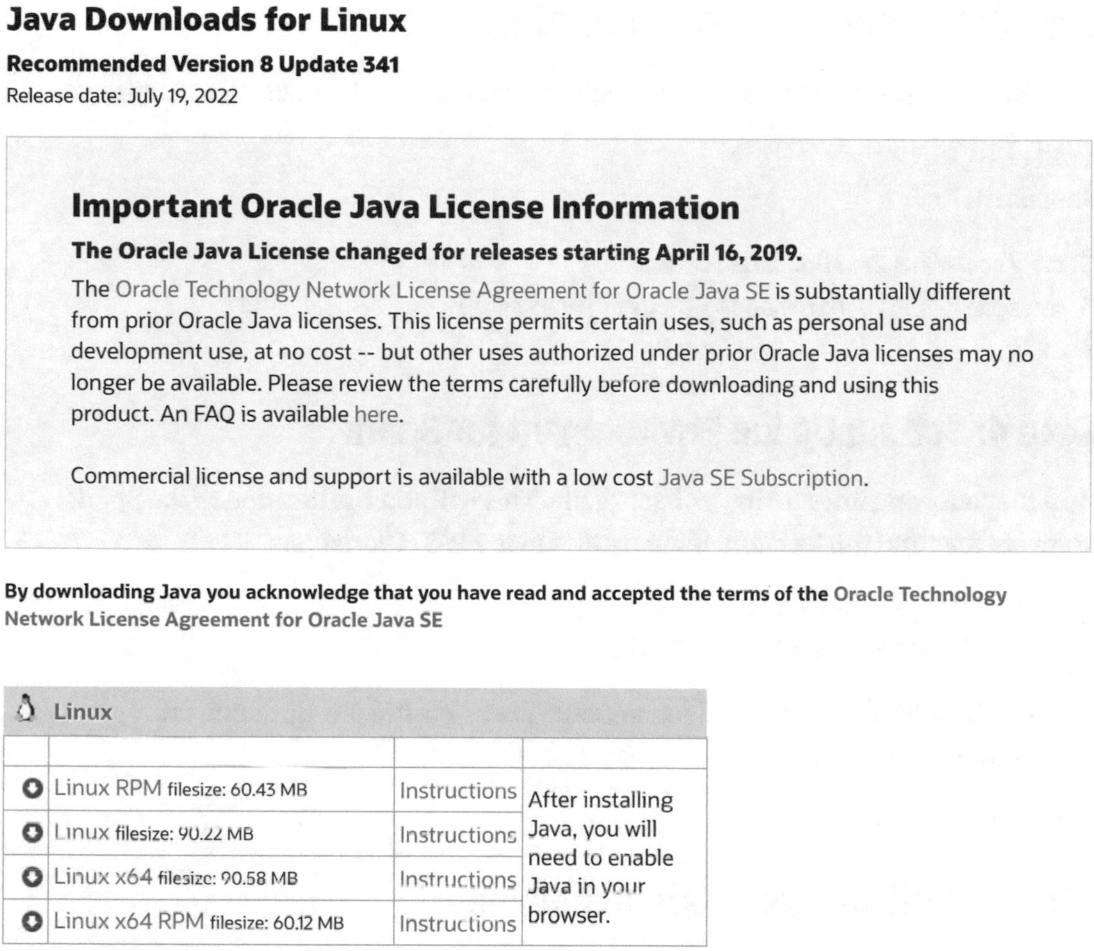

Java Downloads for Linux

Recommended Version 8 Update 341

Release date: July 19, 2022

Important Oracle Java License Information

The Oracle Java License changed for releases starting April 16, 2019.

The Oracle Technology Network License Agreement for Oracle Java SE is substantially different from prior Oracle Java licenses. This license permits certain uses, such as personal use and development use, at no cost -- but other uses authorized under prior Oracle Java licenses may no longer be available. Please review the terms carefully before downloading and using this product. An FAQ is available here.

Commercial license and support is available with a low cost Java SE Subscription.

By downloading Java you acknowledge that you have read and accepted the terms of the Oracle Technology **Network License Agreement for Oracle Java SE**

⌂ Linux		
⊙ Linux RPM filesize: 60.43 MB	Instructions	After installing
⊙ Linux filesize: 90.22 MB	Instructions	Java, you will need to enable
⊙ Linux x64 filesize: 90.58 MB	Instructions	Java in your
⊙ Linux x64 RPM filesize: 60.12 MB	Instructions	browser.

Figure 2-2. *Java download page*

2.   Click the Java file of your choice and save the file to a location (e.g., `/home/<user>/Downloads`).

Step 2: Installing Spark

Extract the Spark .tgz file downloaded before. To unpack the spark-3.3.0-bin-hadoop3.tgz file in Linux, open a terminal window, move to the location in which the file was downloaded

```
$ cd PATH/TO/spark-3.3.0-bin-hadoop3.tgz_location
```

and execute

```
$ tar -xzvf ./spark-3.3.0-bin-hadoop3.tgz
```

Step 3: Moving Spark Software Files

You can move the Spark files to an installation directory such as /usr/local/spark:

```
$ su -
Password:

$ cd /home/<user>/Downloads/
$ mv spark-3.3.0-bin-hadoop3 /usr/local/spark
$ exit
```

Step 4: Setting Up the Environment for Spark

Add the following lines to the ~/.bashrc file. This will add the location of the Spark software files and the location of binary files to the PATH variable:

```
export SPARK_HOME=/usr/local/spark
export PATH=$PATH:$SPARK_HOME/bin
```

Use the following command for sourcing the ~/.bashrc file, updating the environment variables:

```
$ source ~/.bashrc
```

Step 5: Verifying the Spark Installation

Write the following command for opening the Spark shell:

```
$ $SPARK_HOME/bin/spark-shell
```

If Spark is installed successfully, then you will find the following output:

```
Setting default log level to "WARN".
To adjust logging level use sc.setLogLevel(newLevel). For SparkR, use
setLogLevel(newLevel).
22/08/29 22:16:45 WARN NativeCodeLoader: Unable to load native-hadoop
library for your platform... using builtin-java classes where applicable
22/08/29 22:16:46 WARN Utils: Service 'SparkUI' could not bind on port
4040. Attempting port 4041.
Spark context Web UI available at http://192.168.0.16:4041
```

```
Spark context available as 'sc' (master = local[*], app id =
local-1661804206245).
Spark session available as 'spark'.
Welcome to


      ____              __
     / __/__  ___ _____/ /__
    _\ \/ _ \/ _ `/ __/  '_/
   /___/ .__/\_,_/_/ /_/\_\   version 3.3.0
      /_/

Using Scala version 2.12.15 (Java HotSpot(TM) 64-Bit Server VM,
Java 18.0.2)
Type in expressions to have them evaluated.
Type :help for more information.

scala>
```

You can try the installation a bit further by taking advantage of the README.md file that is present in the $SPARK_HOME directory:

```
scala> val readme_file = sc.textFile("/usr/local/spark/README.md")
readme_file: org.apache.spark.rdd.RDD[String] = /usr/local/spark/README.md
MapPartitionsRDD[1] at textFile at <console>:23
```

The Spark context Web UI would be available typing the following URL in your browser:

```
http://localhost:4040
```

There, you can see the jobs, stages, storage space, and executors that are used for your small application. The result can be seen in Figure 2-3.

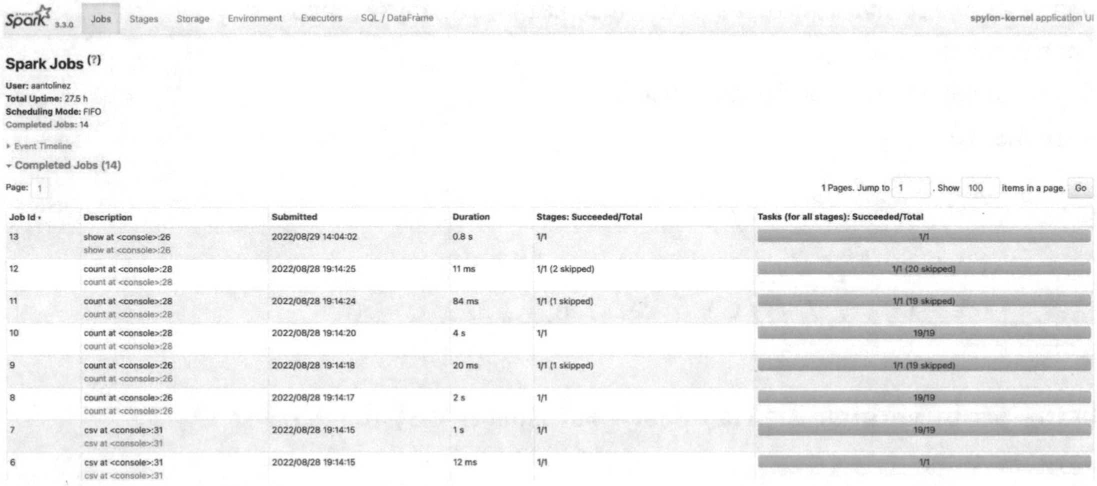

Figure 2-3. *Apache Spark Web UI showing jobs, stages, storage, environment, and executors used for the application running on the Spark shell*

Installation of Apache Spark on Windows

In this section I will show you how to install Apache Spark on Windows 10 and test the installation. It is important to notice that to perform this installation, you must have a user account with administrator privileges. This is mandatory to install the software and modify system PATH.

Step 1: Java Installation

As we did in the previous section, the first step you should take is to be sure you have Java installed and it is accessible by Apache Spark. You can verify Java is installed using the command line by clicking Start, typing cmd, and clicking Command Prompt. Then, type the following command in the command line:

```
java -version
```

If Java is installed, you will receive an output similar to this:

```
openjdk version "18.0.2.1" 2022-08-18
OpenJDK Runtime Environment (build 18.0.2.1+1-1)
OpenJDK 64-Bit Server VM (build 18.0.2.1+1-1, mixed mode, sharing)
```

If a message is instead telling you that command is unknown, Java is not installed or not available. Then you have to proceed with the following steps.

Install Java. In this case we are going to use OpenJDK as a Java Virtual Machine. You have to download the binaries that match your operating system version and hardware. For the purposes of this tutorial, we are going to use OpenJDK JDK 18.0.2.1, so I have downloaded the `openjdk-18.0.2.1_windows-x64_bin.zip` file. You can use other Java distributions as well.

Download the file, save it, and unpack the file in a directory of your choice. You can use any unzip utility to do it.

Step 2: Download Apache Spark

Open a web browser and navigate to the Spark downloads URL and follow the same instructions given in Figure 2-1.

To unpack the `spark-3.3.0-bin-hadoop3.tgz` file, you will need a tool capable of extracting `.tgz` files. You can use a free tool like 7-Zip, for example.

Verify the file integrity. It is always a good practice to confirm the checksum of a downloaded file, to be sure you are working with unmodified, uncorrupted software. In the Spark download page, open the checksum link and copy or remember (if you can) the file's signature. It should be something like this (string not complete):

```
1e8234d0c1d2ab4462 ... a2575c29c     spark-3.3.0-bin-hadoop3.tgz
```

Next, open a command line and enter the following command:

```
certutil -hashfile PATH\TO\spark-3.3.0-bin-hadoop3.tgz SHA512
```

You must see the same signature you copied before; if not, something is wrong. Try to solve it by downloading the file again.

Step 3: Install Apache Spark

Installing Apache Spark is just extracting the downloaded file to the location of your choice, for example, C:\spark or any other.

Step 4: Download the winutils File for Hadoop

Create a folder named Hadoop and a bin subfolder, for example, C:\hadoop\bin, and download the winutils.exe file for the Hadoop 3 version you downloaded before to it.

Step 5: Configure System and Environment Variables

Configuring environment variables in Windows means adding to the system environment and PATH the Spark and Hadoop locations; thus, they become accessible to any application.

You should go to Control Panel ➤ System and Security ➤ System. Then Click "Advanced system settings" as shown in Figure 2-4.

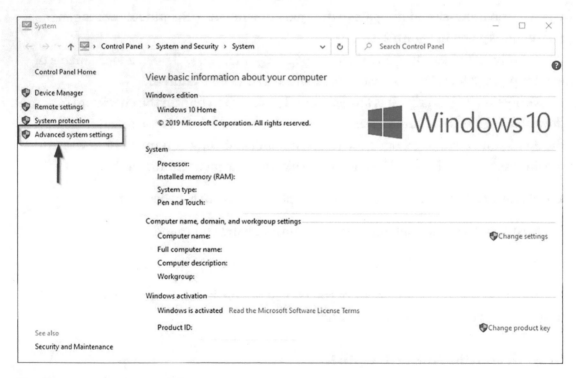

Figure 2-4. *Windows advanced system settings*

You will be prompted with the System Properties dialog box, Figure 2-5 left:

1. Click the Environment Variables button.

2. The Environment Variables window appears, Figure 2-5 top-right:

   a. Click the New button.

3. Insert the following variables:

    a. JAVA_HOME: \PATH\TO\YOUR\JAVA-DIRECTORY

    b. SPARK_HOME: \PATH\TO\YOUR\SPARK-DIRECTORY

    c. HADOOP_HOME: \PATH\TO\YOUR\HADOOP-DIRECTORY

    You will have to repeat the previous step twice, to introduce the
    three variables.

4. Click the OK button to save the changes.

5. Then, click your Edit button, Figure 2-6 left, to edit your PATH.

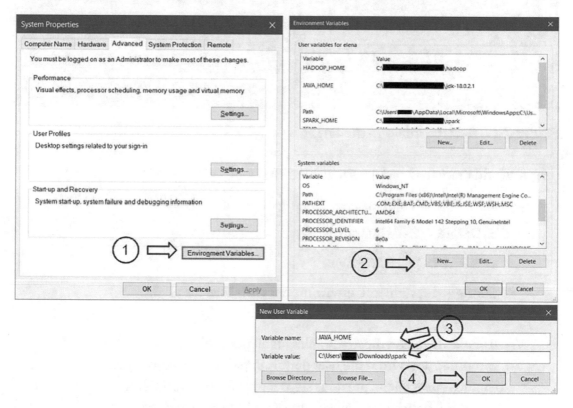

Figure 2-5. *Environment variables*

6.   After that, click the New button, Figure 2-6 right.

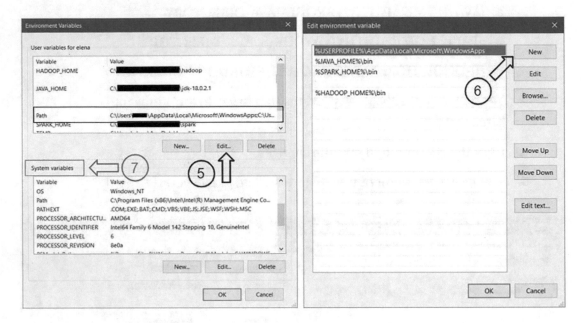

Figure 2-6. *Add variables to the PATH*

And add these new variables to the PATH:

a. %JAVA_HOME%\bin

b. %SPARK_HOME%\bin

c. %HADOOP_HOME%bin

Configure system and environment variables for all the computer's users.

7.   If you want to add those variables for all the users of your computer, apart from the user performing this installation, you should repeat all the previous steps for the System variables, Figure 2-6 bottom-left, clicking the New button to add the environment variables and then clicking the Edit button to add them to the system PATH, as you did before.

Verify the Spark installation.

If you execute spark-shell for the first time, you will see a great bunch of informational messages. This is probably going to distract you and make you fear something went wrong. To avoid these messages and concentrate only on possible error messages, you are going to configure the Spark logging parameters.

Go to your %SPARK_HOME%\conf directory and find a file named log4j2. properties.template, and rename it as log4j2.properties. Open the file with Notepad or another editor.

Find the line

```
rootLogger.level = info
```

And change it as follows:

```
rootLogger.level = ERROR
```

After that, save it and close it.

Open a cmd terminal window and type spark-shell. If the installation went well, you will see something similar to Figure 2-7.

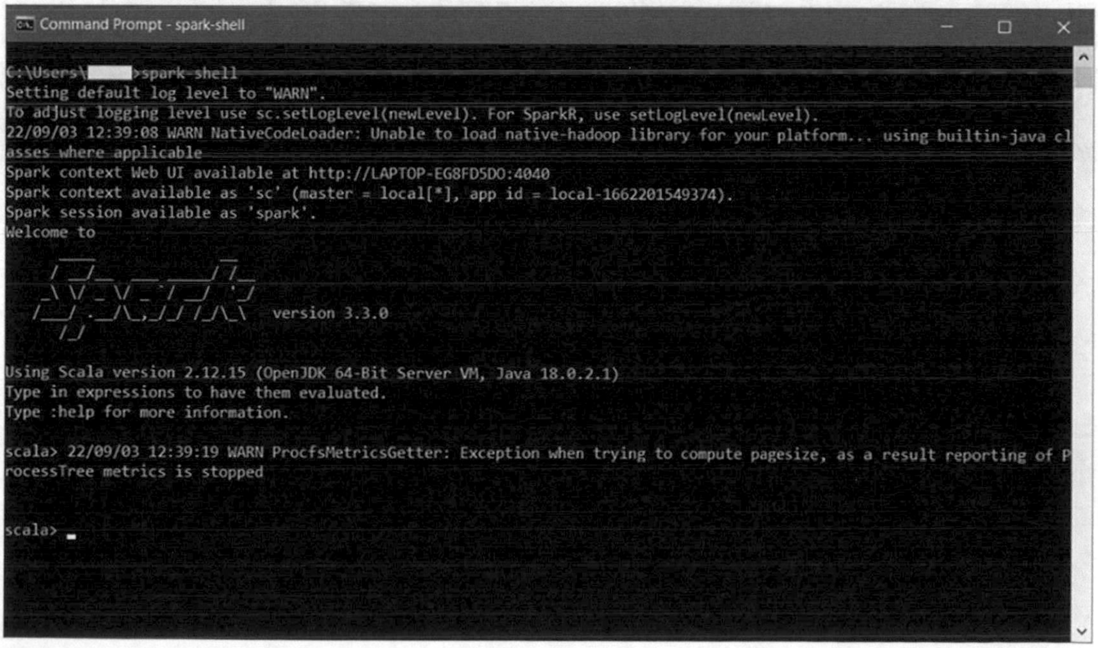

Figure 2-7. *spark-shell window*

To carry out a more complete test of the installation, let's try the following code:

```
val file =sc.textFile("C:\\PATH\\TO\\YOUR\\spark\\README.md")
```

This will create a RDD. You can view the file's content by using the next instruction:

```
file.take(10).foreach(println)
```

You can see the result in Figure 2-8.

Figure 2-8. *spark-shell test code*

To exit spark-shell, you can press Ctrl-D in the Command Prompt window or
type :q.

Open a web browser and type the URL http://localhost:4040/. You can also use the
name of your computer, instead of localhost. You should see an Apache Spark Web UI
similar to the one shown in Figure 2-9. The following example shows the Executors page.

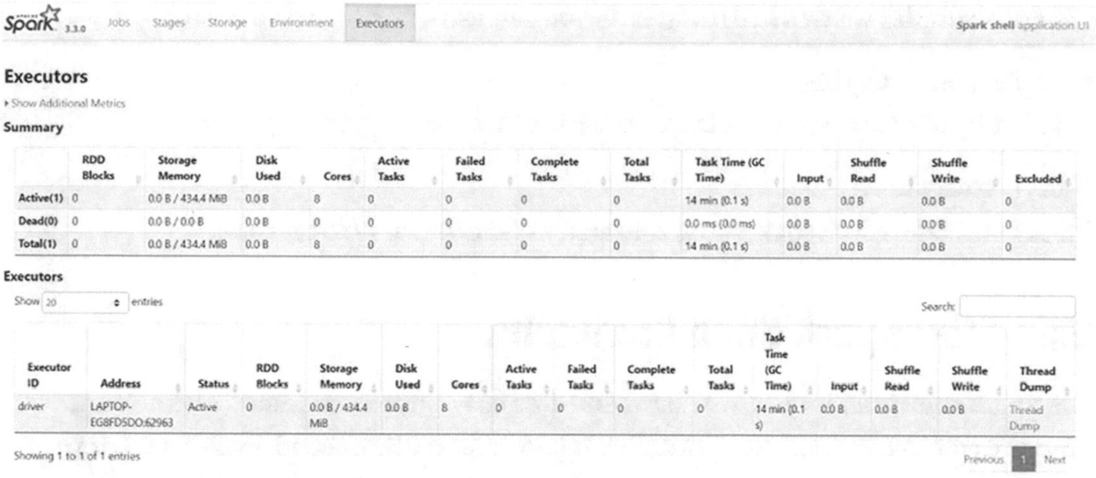

Figure 2-9. *Apache Spark Web UI*

If you have Python installed, you can run PySpark with this command:

```
pyspark
```

The PySpark Command Prompt window can be closed using quit().

2.2 Hands-On Spark Shell

Apache Spark comes with a series of command-line utilities through which you can interact with Spark's APIs. Spark provides shell utilities for several programming languages such as spark-shell for Scala, pyspark for Python, spark-sql for SQL, and sparkR for the R language.

Spark also supplies other specialized command-line tools like spark-submit, run-example, and spark-class. You can use spark-submit to execute self-contained applications written in Java, Scala, Python, or R using the Spark API and submit them to the different Spark cluster managers (YARN, Kubernetes, Mesos, and Standalone), supplying execution time options and configurations. Apache Spark comes with several examples coded in Scala, Java, Python, and R, which are located in the examples directory. The shell spark-example can be used to run examples written in Scala and Java:

```
#For Scala and Java examples:
$ $SPARK_HOME/bin/run-example SparkPi
```

For examples written in Python or R, you can use spark-submit directly:

```
#For Python examples:
$ $SPARK_HOME/bin/spark-submit examples/src/main/python/pi.py

#For R examples:
$ $SPARK_HOME/bin/spark-submit examples/src/main/r/dataframe.R
```

Using the Spark Shell Command

The Spark shell is an interactive command-line environment to interact with Spark from the command line. The Spark shell is known as a REPL (Read-Eval-Print Loop) shell interface. A REPL interface reads each input line, evaluates it, and returns the result. It is mostly used to run ad hoc queries against a Spark cluster. The Spark shell is a very convenient tool to debug your software and explore new features while getting immediate feedback. There are specific shell scripts for different languages such as spark-shell to launch the Spark Scala shell; pyspark for Spark with Python, also called PySpark; and sparkr to submit R language programs to Spark.

The Scala Shell Command Line

One of the main features of the Spark shell is that it creates the SparkSession and SparkContext for you. You can access the SparkSession and SparkContext through their objects, spark for the former and sc for the latter.

Remember you can start the spark-shell script by executing

```
$SPARK_HOME/bin/spark-shell
```

```
~ % spark-shell
Setting default log level to "WARN".
To adjust logging level use sc.setLogLevel(newLevel). For SparkR, use
setLogLevel(newLevel).
Spark context Web UI available at http://192.168.0.16:4040
Spark context available as 'sc' (master = local[*], app id =
local-1662487353802).
Spark session available as 'spark'.
Welcome to
```

```
      ___              __
     / __/__  ___ ___ / /__
    _\ \/ _ \/ _ `/ _/  '_/
   /___/ .__/\_,_/ /_/_/\_\   version 3.3.0
      /_/
```

Using Scala version 2.12.15 (Java HotSpot(TM) 64-Bit Server VM,
Java 18.0.2)
Type in expressions to have them evaluated.
Type :help for more information.

```
scala>
```

This automatically instantiates the SparkSession as spark and SparkContext as sc:

```
scala> :type spark
org.apache.spark.sql.SparkSession

scala> :type sc
org.apache.spark.SparkContext
```

You can access environment variables from the shell using the getenv method as System.getenv('ENV_NAME'), for example:

```
scala> System.getenv("PWD")
res12: String = /Users/aantolinez
```

spark-shell also provides online help by typing

```
scala> :help
All commands can be abbreviated, e.g., :he instead of :help.
:completions <string> output completions for the given string
:edit <id>|<line>     edit history
:help [command]       print this summary or command-specific help
:history [num]        show the history (optional num is commands to show)
:h? <string>          search the history
...
...
:save <path>          save replayable session to a file
:settings <options>   update compiler options, if possible; see reset
```

```
:silent               disable/enable automatic printing of results
:warnings             show the suppressed warnings from the most recent
line which had any

scala>
```

Please, do not confuse the inline help provided by :help with spark-shell runtime options shown by the spark-shell -h option.

The -h option permits passing runtime environment options to the shell, allowing a flexible execution of your application, depending on the cluster configuration. Let's see some examples in which we run Apache Spark with Apache Hudi; set the cluster manager (YARN), the deployment mode, and the number of cores per executor; and allocate the memory available for the driver and executors:

```
$SPARK_HOME/bin/spark-shell \
--master yarn \
--deploy-mode cluster \
--driver-memory 16g \
--executor-memory 32g \
--executor-cores 4   \
--conf "spark.sql.shuffle.partitions=1000" \
--conf "spark.executor.memoryOverhead=4024" \
--conf "spark.memory.fraction=0.7" \
--conf "spark.memory.storageFraction=0.3" \
--packages org.apache.hudi:hudi-spark3.3-bundle_2.12:0.12.0 \
--conf "spark.serializer=org.apache.spark.serializer.KryoSerializer" \
--conf "spark.sql.catalog.spark_catalog=org.apache.spark.sql.hudi.catalog.
HoodieCatalog" \
--conf "spark.sql.extensions=org.apache.spark.sql.hudi.
HoodieSparkSessionExtension"
```

In the next example, we define at runtime the database driver and version we would like to be used:

```
$SPARK_HOME/bin/spark-shell \
--master yarn \
--deploy-mode cluster \
--driver-memory 16g \
```

```
--executor-memory 32g \
--executor-cores 4   \
--driver-class-path /path/to/postgresql-42.5.0.jar \
--conf "spark.sql.shuffle.partitions=1000" \
--conf "spark.executor.memoryOverhead=4024" \
--conf "spark.memory.fraction=0.7" \
--conf "spark.memory.storageFraction=0.3" \
```

Another important spark-shell feature is that, by default, it starts using local[*] as master and assigns a spark.app.id with local-xxx schema and a spark.app.name="Spark shell". All the properties set by default and environment variables used in the running environment can be examined by looking at the Web UI launched by spark-shell and accessible via the URL http://localhost:4040/environment/ as can be seen in Figure 2-10.

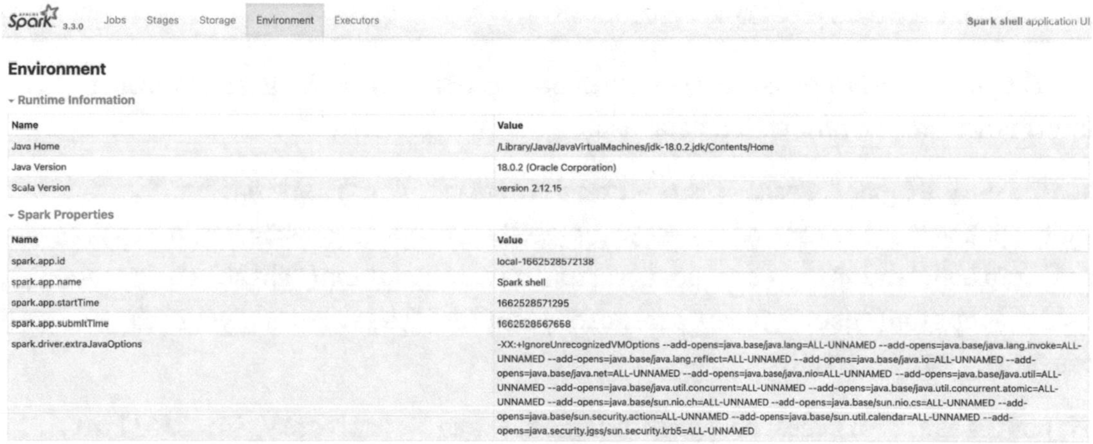

Figure 2-10. *The Apache Spark Web UI on port 4040*

As mentioned before, you can run interactive applications by typing your code in the command line. For example, let's create a dataframe from a data sequence:

```scala
scala> import spark.implicits._
import spark.implicits._

scala> val cars=Seq(("USA","Chrysler","Dodge","Jeep"),("Germany","BMW","VW",
"Mercedes"),("Spain", "GTA Spano","SEAT","Hispano Suiza"))
```

```
cars: Seq[(String, String, String, String)] =
List((USA,Chrysler,Dodge,Jeep), (Germany,BMW,VW,Mercedes), (Spain,GTA
Spano,SEAT,Hispano Suiza))

scala> val cars_df = cars.toDF()
cars_df: org.apache.spark.sql.DataFrame = [_1: string, _2: string ... 2
more fields]

scala> cars_df.show()
+-------+---------+-----+-------------+
|     _1|       _2|   _3|           _4|
+-------+---------+-----+-------------+
|    USA| Chrysler|Dodge|         Jeep|
|Germany|      BMW|   VW|     Mercedes|
|  Spain|GTA Spano| SEAT|Hispano Suiza|
+-------+---------+-----+-------------+
```

The following is a practical example of how spark-shell can be used to retrieve information from a production database:

```
scala> val df_postgresql = spark.read.format("jdbc").option("url",
"jdbc:postgresql://ec2-52-77-8-54.us-west-1.compute.amazonaws.com:5432/
db").option("driver", "org.postgresql.Driver").option("dbtable","schema.
table").option("user","user_password").option("password", "your_db_
password_here").load()

df_postgresql: org.apache.spark.sql.DataFrame = [category_id: smallint,
category_name: string ... 2 more fields]

scala> df_postgresql.show()
+-----------+--------------+--------------------+-------+
|category_id| category_name|         description|picture|
+-----------+--------------+--------------------+-------+
|          1|     Beverages|Soft drinks, coff...|     []|
|          2|    Condiments|Sweet and savory ...|     []|
|          3|   Confections|Desserts, candies...|     []|
|          4|Dairy Products|             Cheeses|     []|
```

```
|          5|Grains/Cereals|Breads, crackers,...|      []|
|          6|  Meat/Poultry|     Prepared meats|      []|
|          7|       Produce|Dried fruit and b...|      []|
|          8|       Seafood|   Seaweed and fish|      []|
+-----------+--------------+--------------------+------+
```

Finally, to leave spark-shell, you just need to type :q.

The Pyspark Shell Command Line

If you prefer programming with Python, you can invoke the pyspark shell. As with spark-shell, you can run it by typing in a terminal:

$SPARK_HOME/bin/pyspark

```
~ % pyspark
Python 3.6.12 (default, May 18 2021, 22:47:55)
[GCC 4.8.5 20150623 (Red Hat 4.8.5-28)] on linux
Type "help", "copyright", "credits" or "license" for more information.
SLF4J: Class path contains multiple SLF4J bindings.
SLF4J: Found binding in [jar:file:/usr/share/aws/glue/etl/jars/glue-
assembly.jar!/org/slf4j/impl/StaticLoggerBinder.class]
...
...
22/09/07 16:30:26 WARN Client: Same path resource file:///usr/share/aws/
glue/libs/pyspark.zip added multiple times to distributed cache.
Welcome to
      ____              __
     / __/__  ___ _____/ /__
    _\ \/ _ \/ _ `/ __/  '_/
   /__ / .__/\_,_/_/ /_/\_\   version 3.3.0
      /_/

Using Python version 3.6.12 (default, May 18 2021 22:47:55)
SparkSession available as 'spark'.
>>>
```

As with `spark-shell`, the pyspark shell automatically creates a SparkSession accessible as `spark` and a SparkContext available through the variable `sc`. It also sets up a shell Web UI server with URL http://localhost:4040/. By default, it uses port 4040; however, if this port couldn't bind, ports 4041, 4042, and so on will be explored until one is found that binds.

As mentioned before with `spark-shell`, you can run interactive applications in `spark` by typing your code in the command line. In the following example, we create a dataframe from a list of tuples:

```
>> cars = [("USA","Chrysler","Dodge","Jeep"),("Germany","BMW","VW",
"Mercedes"),("Spain", "GTA Spano","SEAT","Hispano Suiza")]
>> cars_df = spark.createDataFrame(cars)
>> cars_df.show()
+-------+---------+-----+-------------+
|     _1|       _2|   _3|           _4|
+-------+---------+-----+-------------+
|    USA| Chrysler|Dodge|         Jeep|
|Germany|      BMW|   VW|     Mercedes|
|  Spain|GTA Spano| SEAT|Hispano Suiza|
+-------+---------+-----+-------------+
```

The pyspark shell also admits runtime configuration parameters. The following is an example of pyspark executed with configuration options:

```
$SPARK_HOME/bin/pyspark \
--master yarn \
--deploy-mode client \
--executor-memory 16G \
--executor-cores 8 \
--conf spark.sql.parquet.mergeSchema=true \
--conf spark.sql.parquet.filterPushdown=true \
--conf spark.sql.parquet.writeLegacyFormat=false Hands_On_Spark3_Script.py
> ./Hands_On_Spark3_Script.log 2>&1 &
```

Unlike spark-shell, to leave pyspark, you can type in your terminal one of the commands `quit()` and `exit()` or press Ctrl-D.

Running Self-Contained Applications with the `spark-submit` Command

If you look at the $SPARK_HOME/bin directory, you will find several `spark-submit` scripts. Spark provides different scripts for distinct operating systems and purposes. Therefore, for Linux and Mac, you need the `spark-submit.sh` script file, while for Windows you have to employ the `spark-submit.cmd` command file.

The spark-submit command usage is as follows:

```
~ % $SPARK_HOME/bin/spark-submit --help

Usage: spark-submit [options] <app jar | python file | R file> [app
arguments]
Usage: spark-submit --kill [submission ID] --master [spark://...]
Usage: spark-submit --status [submission ID] --master [spark://...]
Usage: spark-submit run-example [options] example-class [example args]
```

Some of the most common spark-submit options are

```
$ $SPARK_HOME/bin/spark-submit \
  --master <master-url> \
  --deploy-mode <deploy-mode> \
  --class <main-class> \
  --conf <key>=<value> \
  --driver-memory <value>g \
  --executor-memory <value>g \
  --executor-cores <number of cores>  \
  --jars  <comma separated dependencies>
  ... # other options
  <application-jar> \
  [application-arguments]
```

Now, we are going to illustrate how spark-submit works with a practical example:

```
$ $SPARK_HOME/bin/spark-submit \
--deploy-mode client \
--master local \
--class org.apache.spark.examples.SparkPi \
/$SPARK_HOME/examples/jars/spark-examples_2.12-3.3.0.jar 80
```

If you have run this piece of code as it is, you would have surely seen some stuff like this coming out to your console:

```
22/08/30 22:04:30 WARN Utils: Your hostname, MacBook-Pro.local resolves
to a loopback address: 127.0.0.1; using 192.168.0.16 instead (on
interface en0)
22/08/30 22:04:30 WARN Utils: Set SPARK_LOCAL_IP if you need to bind to
another address
22/08/30 22:05:00 INFO SparkContext: Running Spark version 3.3.0
. . .
. . .
22/08/30 22:05:03 INFO DAGScheduler: Job 0 finished: reduce at SparkPi.
scala:38, took 1.174952 s
```
Pi is roughly 3.142520392815049
```
22/08/30 22:05:03 INFO SparkUI: Stopped Spark web UI at
http://192.168.0.16:4043
22/08/30 22:05:03 INFO MapOutputTrackerMasterEndpoint:
MapOutputTrackerMasterEndpoint stopped!
22/08/30 22:05:03 INFO MemoryStore: MemoryStore cleared
22/08/30 22:05:03 INFO BlockManager: BlockManager stopped
22/08/30 22:05:03 INFO BlockManagerMaster: BlockManagerMaster stopped
22/08/30 22:05:03 INFO OutputCommitCoordinator$OutputCommitCoordinatorEndpo
int: OutputCommitCoordinator stopped!
22/08/30 22:05:03 INFO SparkContext: Successfully stopped SparkContext
22/08/30 22:05:03 INFO ShutdownHookManager: Shutdown hook called
22/08/30 22:05:03 INFO ShutdownHookManager: Deleting directory /
private/var/folders/qd/6ly2_9_54tq434fctwmmsc3m0000gp/T/spark-30934777-
e061-403c-821d-0bbaa2e62745
22/08/30 22:05:03 INFO ShutdownHookManager: Deleting directory /private/
var/folders/qd/6ly2_9_54tq434fctwmmsc3m0000gp/T/spark-3e2e234c-7615-4a9c-
b34f-878d149517e2
```

"**Pi is roughly 3.142520392815049**" in bold is the information you are really interested in.

As we mentioned before in "Step 5: Configure System and Environment Variables," to avoid such a bunch of information that makes it difficult to find the final outcome of your application, you can tune your Spark configuration a little bit to show you just ERROR messages, removing INFO lines.

Assuming you are using Apache Spark 3.3.0, in a terminal window, do the following:

```
$ ls $SPARK_HOME/conf
fairscheduler.xml.template        spark-defaults.conf.template
log4j2.properties.template        spark-env.sh.template
metrics.properties.template        workers.template
```

Rename the log4j2.properties.template file, and name it as log4j2.properties:

```
$ mv $SPARK_HOME/conf/log4j2.properties.template $SPARK_HOME/conf/log4j2.
properties
```

Edit the log4j2.properties file:

```
$ vi $SPARK_HOME/conf/log4j2.properties
```

Find the following line

```
# Set everything to be logged to the console
rootLogger.level = info
```

And change it to

```
# Set everything to be logged to the console
rootLogger.level = ERROR
```

Save the file, and run the Spark example application again:

```
$ $SPARK_HOME/bin/spark-submit \
--name "Hands-On Spark 3" \
--master local\[4] \
--deploy-mode client \
--conf spark.eventLog.enabled=false \
--conf "spark.executor.extraJavaOptions=-XX:+PrintGCDetails
-XX:+PrintGCTimeStamps" \
```

```
--class org.apache.spark.examples.SparkPi \
/$SPARK_HOME/examples/jars/spark-examples_2.12-3.3.0.jar 80
```

Pi is roughly 3.1410188926273617
$

This time you will see a clean exit.

Let's now review the meaning of the most common options.

Spark Submit Options

You can get all `spark-submit` options available by running the following command:

```
$ $SPARK_HOME/bin/spark-submit --help
```

Let's explain the most relevant `spark-submit` options and configurations used with Scala and Python—PySpark.

Deployment Mode Options

The `--deploy-mode` option specifies whether the driver program will be launched locally ("client") or will be run in a cluster ("cluster"). Table 2-1 describes the meaning of each option.

Table 2-1. *The Apache Spark Deployment Modes*

Option	Description
cluster	In cluster mode, the driver program will run in one of the worker machines inside a cluster. Cluster mode is used to run production jobs.
client (default option)	In client mode, the driver program runs locally where the application is submitted and the executors run in different nodes.

Cluster Manager Options

The `--master` option specifies what cluster manager to use and the master URL for the cluster to run your application in. You can see the different cluster managers available and how to use them in Table 2-2.

Table 2-2. *The Apache Spark Cluster Managers*

Option	Template	Description
Standalone	spark://IP:PORT	URL of the master node, IP address, and port, which is 7077 by default.
Mesos	mesos://host:port	The master URLs for Mesos.
YARN	--master yarn	Cluster resources managed by Hadoop YARN.
Kubernetes	k8s://https://host:port	Kubernetes host and port: k8s://https://<k8s-apiserver-host>:<k8s-apiserver-port>
Local	local	Run Spark locally with no parallelism and just one worker thread (i.e., at all).
	local[K],	Run Spark locally with K number of cores or worker threads.
	local[K,F]	Run Spark locally with K worker threads and F maxFailures or number of attempts it should try when failed.
	local[*] (Default: local[*])	Run Spark locally with as many worker threads as logical cores are available.

Here are a few examples of these common options:

```
# Run application locally on 4 cores
(base) aantolinez@MacBook-Pro ~ % $SPARK_HOME/bin/spark-submit \
 --class org.apache.spark.examples.SparkPi \
 --master local[4] \
 /$SPARK_HOME/examples/jars/spark-examples_2.12-3.3.0.jar 80

zsh: no matches found: local[4]
```

Note   If you are using the zsh, also called the Z shell, you would have to escape [4] in --master local[4] \ with "\", like --master local\[4] \; otherwise, you will get the following error message:

```
zsh: no matches found: local[4]
```

```
(base) aantolinez@MacBook-Pro ~ % $SPARK_HOME/bin/spark-submit \
 --class org.apache.spark.examples.SparkPi \
 --master local\[4] \
 /$SPARK_HOME/examples/jars/spark-examples_2.12-3.3.0.jar 80

# Spark standalone cluster and a Python application
$SPARK_HOME/bin/spark-submit \
  --master spark://192.168.1.3:7077 \
/$SPARK_HOME/examples/src/main/python/pi.py

# Spark standalone cluster in client deploy mode and 100 cores
$SPARK_HOME/bin/spark-submit \
  --class org.apache.spark.examples.SparkPi \
  --master spark://192.168.1.3:7077 \
  --executor-memory 20G \
  --total-executor-cores 100 \
/$SPARK_HOME/examples/jars/spark-examples_2.12-3.3.0.jar 80

# Spark standalone cluster in cluster deploy mode with supervised option.
# --supervise automatically restarts the driver if it fails with a non-zero
exit code.
$SPARK_HOME/bin/spark-submit \
  --class org.apache.spark.examples.SparkPi \
  --master spark://192.168.1.3:7077 \
  --deploy-mode cluster \
  --supervise \
  --executor-memory 20G \
  --total-executor-cores 100 \
  /$SPARK_HOME/examples/jars/spark-examples_2.12-3.3.0.jar 80

# Spark on a YARN cluster in cluster deploy mode
export HADOOP_CONF_DIR=PATH_TO_HADOOP_CONF_DIR
$SPARK_HOME/bin/spark-submit \
  --master yarn \
  --deploy-mode cluster \
  --class org.apache.spark.examples.SparkPi \
  --executor-memory 10G \
```

```
  --num-executors 20 \
  /$SPARK_HOME/examples/jars/spark-examples_2.12-3.3.0.jar 80

# Run on a Mesos cluster in cluster deploy mode
$SPARK_HOME/bin/spark-submit \
  --class org.apache.spark.examples.SparkPi \
  --master mesos://207.184.161.138:7077 \
  --deploy-mode cluster \
  --executor-memory 16G \
  --total-executor-cores 64 \
  /$SPARK_HOME/examples/jars/spark-examples_2.12-3.3.0.jar 80

# Spark on a Kubernetes cluster in cluster deploy mode
$SPARK_HOME/bin/spark-submit \
  --class org.apache.spark.examples.SparkPi \
  --master k8s://xx.yy.zz.ww:443 \
  --deploy-mode cluster \
  --executor-memory 20G \
  --num-executors 50 \
  /$SPARK_HOME/examples/jars/spark-examples_2.12-3.3.0.jar 80
```

Tuning Resource Allocation

When submitting an application to your cluster, you can pay attention to its execution performance, thus making your Spark program run faster. The two main resources Apache Spark cares about are CPU and RAM. To take advantage of what a Spark cluster can offer, you can control how much memory and cores the driver and executors can use.

The options enumerated in this section not only affect Spark performance but the cluster in which it could be running as well. For example, they influence how the resources requested by Spark will fit into what the cluster manager has available.

In Table 2-3 you can see which parameters to tune and their technical description.

Table 2-3. *The Apache Spark Driver and Executor Resource Management Options*

Option	Description
--executor-cores	• Number of CPU cores to be used by the Spark driver for the executor process. • The cores property controls the number of concurrent tasks an executor can run. • --executor-cores 5 means that each executor can run a maximum of five tasks at the same time.
--executor-memory	• Amount of RAM to use for the executor process. • This option affects the maximum size of data Spark can cache and allocate for shuffle data structures. • This property impacts operations performance like aggregations, grouping, and joins.
--num-executors (*)	It controls the number of executors requested.
--driver-memory	Memory to be used by the Spark driver.
--driver-cores	The number of CPU cores given to the Spark driver.
--total-executor-cores	The total number of cores granted to the executor.

(*) Note   Starting with CDH 5.4/Spark 1.3, you can bypass setting up this parameter with the spark.dynamicAllocation.enabled property, turning on dynamic allocation. Dynamic allocation permits your application to solicit available executors while there are pending tasks and release them when unused.

Dynamically Loading Spark Submit Configurations

In general it is recommended to avoid hard-coding configurations in your application using the SparkConf if your application could be run in different cluster configurations such as different cluster managers, distinct amounts of memory available, etc. because it cannot be modified by the user once a SparkConf object has been passed to Apache Spark.

Thus, instead of establishing SparkContext configuration in your source code, as you can see in the following code snippet, it is better to leave that configuration ready to receive dynamic parameters when the program is called:

```scala
// Hard-coding cluster configuration parameters in Scala
// Create Spark configuration
val conf = new SparkConf()
          .setMaster("local[4]")
          .setAppName("Hands-On Spark 3")
          .set("spark.executor.memory", "32g")
          .set("spark.driver.memory", "16g")
// Create Spark context
val sc = new SparkContext(conf)
```

```python
#Hard-coding cluster configuration parameters in PySpark
conf = SparkConf()
conf.setMaster("spark://localhost:7077")
conf.setAppName("Hands-On Spark 3")
conf.set("spark.executor.memory", "32g")
conf.set("spark.driver.memory", "16g")

sc = SparkContext(conf=conf)
```

The SparkContext is created only once for an application; thus, another more flexible approach to the problem could be constructing it with a void configuration:

```scala
// SparkContext with a void configuration in Scala
val sc = new SparkContext(new SparkConf())
```

```python
# SparkContext with a void configuration in PySpark
conf = SparkConf()
sc = SparkContext(conf=conf)
```

Then you can dynamically pass configuration parameters to your cluster at runtime:

```
$SPARK_HOME/bin/spark-submit \
--name "Hands-On Spark 3" \
--master local[4] \
--deploy-mode client \
```

```
--conf spark.eventLog.enabled=false \
--conf "spark.executor.extraJavaOptions=-XX:+PrintGCDetails
-XX:+PrintGCTimeStamps" \
--class org.apache.spark.examples.SparkPi \
/$SPARK_HOME/examples/jars/spark-examples_2.12-3.3.0.jar 80
```

Spark submit allows you to fine-tune your cluster configuration with dozens of parameters that can be sent to the SparkContext using the --config/-c option or by setting the SparkConf to create a SparkSession.

These options control application properties (Table 2-4), the runtime environment (Table 2-5), shuffle behavior, Spark UI, compression and serialization, memory management, execution behavior, executor metrics, networking, scheduling, barrier execution mode, dynamic allocation (Table 2-6), thread configurations, security, and runtime SQL configuration, among others (Table 2-7). Next, we will explore some of the most common ones.

Application Properties

Table 2-4. *Spark application properties*

Property	Description
spark.app.name (Default value, none)	The name of your application.
spark.driver.cores (Default value, 1)	In cluster mode, the number of cores to use for the driver process only.
spark.driver.memory (Default value, 1g)	Amount of memory to use for the driver process. In client mode, it should be set via the --driver-memory command-line option or the properties file.

Runtime Environment

Table 2-5. *Spark runtime environment*

Property	Description

`spark.driver.`
`extraClassPath`
(Default value, none)
	Extra classpath entries to prepend to the classpath of the driver. In client mode, it should be set via the `--driver-class-path` command-line option or in the default properties file.
	The option allows you to load specific JAR files, such as database connectors and others.

Dynamic Allocation

Table 2-6. *Spark allocation resources*

Property	Description

`spark.dynamicAllocation.`
`enabled`
(Default value, false)
	Whether to use dynamic resource allocation to adjust the number of executors processing your application based on existing workload.

`spark.dynamicAllocation.`
`executorIdleTimeout`
(Default value, 60 s)
	If dynamic allocation is enabled, an executor process will be killed if it has been idle for a longer time.

`spark.dynamicAllocation.`
`cachedExecutorIdleTimeout`
(Default value, infinity)
	If dynamic allocation is enabled, an executor process will be killed if it has cached data and has been idle for a longer time.

Others

Table 2-7. *Other Spark options to control application properties*

Property	Description
`spark.sql.shuffle.partitions` (Default value, 200)	Number of partitions to use when shuffling data for joins or aggregations.
`spark.rdd.compress` (Default value, false)	Whether to compress serialized RDD partitions saving considerable space at the cost of extra CPU processing time.
`spark.executor.pyspark.memory` (Default value, not set)	If set, the amount of memory to be allocated to PySpark in each executor. This option has different behaviors depending on the operating system.
`spark.executor.memoryOverhead` (`executorMemory * spark.executor.memoryOverheadFactor`), minimum of 384	Amount of additional memory allocated per executor process. The maximum memory to run per executor is determined by the sum of `spark.executor.memoryOverhead`, `spark.executor.memory`, `spark.memory.offHeap.size`, and `spark.executor.pyspark.memory`.

The following is an example of the use of some of these options in the command line:

```
$SPARK_HOME/bin/spark-submit \
--master yarn \
--deploy-mode cluster \
--conf "spark.sql.shuffle.partitions=10000" \
--conf "spark.executor.memoryOverhead=8192" \
--conf "spark.memory.fraction=0.7" \
--conf "spark.memory.storageFraction=0.3" \
--conf "spark.dynamicAllocation.minExecutors=10" \
--conf "spark.dynamicAllocation.maxExecutors=2000" \
--conf "spark.dynamicAllocation.enabled=true" \
--conf "spark.executor.extraJavaOptions=-XX:+PrintGCDetails
-XX:+PrintGCTimeStamps" \
```

```
--files /path/of/config.conf, /path/to/mypropeties.json \
--class org.apache.spark.examples.SparkPi \
/$SPARK_HOME/examples/jars/spark-examples_2.12-3.3.0.jar 80
```

Configuration can be passed to Spark in three different ways. Hard-coding, taking advantage of SparkConf properties, overrides others, taking the first order of priority:

```
val config = new SparkConf()
config.set("spark.sql.shuffle.partitions","500")
val spark=SparkSession.builder().appName("Hands-On Spark 3").config(config)
```

The second order of priority would be via `spark-submit`, as part of its --config attributes. And finally, the last one is through the `$SPARK_HOME/conf/spark-defaults.conf` file. The last has the advantage that the configuration established in this file applies globally, meaning to all Spark applications running in the cluster.

In the following, we show an example of a user-developed application and how it can be submitted to Apache Spark.

Let's develop a small program called Functions, which performs a single operation, adding two numbers:

```
object Functions {
    def main(args: Array[String]) = {
        agregar(1,2)
    }
    val agregar = (x: Int, y: Int) => println(x+y)
}
```

Save the file as Functions.scala and compile it as follows. We assume you have Scala installed on your computer:

```
~ % scalac ./Functions.scala -d Functions.jar
```

Then, submit the JAR file to your Spark installation:

```
 ~ % spark-submit --class Functions ./Functions.jar
3
```

You can see the number 3 as program output.

2.3 Spark Application Concepts

Hitherto, you have downloaded and configured Apache Spark, gotten familiar with the Spark shell, and executed some small examples. Now, let's review some important terminology of the Spark application necessary to understand what is happening when you execute your code:

- ***Spark application*** is a user-developed program on top of Spark that uses its APIs and consists of an executor running on a cluster and a driver program.

- ***SparkSession*** is the entry point to communicate with Spark and allows user interaction with the underlying functionalities through Spark APIs.

- ***Tasks*** are the smallest execution unit and are executed inside an executor.

- ***Stages*** are collections of tasks running the same code, each of them in different chunks of a dataset. The use of functions that require a shuffle or reading a dataset, such as reduceByKey(), Join() etc., will trigger in Spark the creation of a stage.

- ***Jobs*** comprise several stages and can be defined as entities that permit the execution and supervision of applications in a Spark cluster.

Spark Application and SparkSession

As we have mentioned before, the SparkSession is the entry point to communicate with Spark and to have access to its functionalities available via the Spark Dataset and DataFrame APIs. The SparkSession is created using the `SparkSession.builder()` constructor, and creating a SparkSession is the first statement in an application.

The SparkSession was introduced with Spark 2.0, and the new class `org.apache.spark.sql.SparkSession` was provided to replace `SparkContext`, `SQLContext`, `StreamingContext`, and `HiveContext`, contexts available prior to version 2.0.

The number of SparkSessions is pretty much unbounded; it means that you can have as many SparkSessions as needed. This is particularly useful when several programmers are working at the same time on the same cluster or when you want to logically segregate your Spark relational entities. However, you can only have one SparkContext on a single JVM. New SparkSessions can be created using either SparkSession.builder() or SparkSession.newSession().

Access the Existing SparkSession

In environments that have been created up front, if necessary, you can access the existing SparkSession from your application using the SparkSession.Builder class with the method getOrCreate() to retrieve an existing session. In the following there are two code snippets in Scala and PySpark showing you how to do it:

```
// Know about the existing SparkSession in Scala
import org.apache.spark.sql.SparkSession
val currentSparkSession = SparkSession.builder().getOrCreate()
print(currentSparkSession)

// SparkSession output
org.apache.spark.sql.SparkSession@7dabc2f9

# Know about the existing SparkSession in PySpark
currentSparkSession = SparkSession.builder.getOrCreate
print(currentSparkSession)

# SparkSession output
pyspark.sql.session.SparkSession object at 0x7fea951495e0
```

You can get the active SparkSession for the current thread, returned by the builder, using the getActiveSession() method:

```
# The active SparkSession for the current thread
s = SparkSession.getActiveSession()
print(s)
```

You can also create a new/another SparkSession using the newSession() method. This method will create a new session with the same app name, master mode, and SparkContext of the active session. Remember that you can have one context for each Spark application:

```
// Create a new SparkSession
val aNewSession = spark.newSession()
print(aNewSession)
org.apache.spark.sql.SparkSession@2dc9b758
```

Out[6]: aNewSession: org.apache.spark.sql.SparkSession = org.apache.spark.
sql.SparkSession@2dc9b758

Get the Current Spark Context Settings/Configurations

Spark has a certain number of settings and configurations you might be interested
in specifying, including application properties and runtime parameters. You can use the
following code in Scala or PySpark to collect all the current configurations:

```
// Get all Spark Configs
val configMap:Map[String, String] = spark.conf.getAll
```

The output you will receive could be similar to this:

```
configMap: Map[String,String] = Map(spark.sql.warehouse.dir -> file:/
Users/.../spark-warehouse, spark.executor.extraJavaOptions -> -XX:+Ignore
UnrecognizedVMOptions --add-opens=java.base/java.lang=ALL-UNNAMED --add-
opens=java.base/java.lang.invoke=ALL-UNNAMED --add-opens=java.base/java.
lang.reflect=ALL-UNNAMED --add-opens=java.base/java.io=ALL-UNNAMED --add-
opens=java.base/java.net=ALL-UNNAMED --add-opens=java.base/java.nio=ALL-
UNNAMED --add-opens=java.base/java.util=ALL-UNNAMED --add-opens=java.
base/java.util.concurrent=ALL-UNNAMED --add-opens=java.base/java.util.
concurrent.atomic=ALL-UNNAMED --add-opens=java.base/sun.nio.ch=ALL-UNNAMED
--add-opens=java.base/sun.nio.cs=ALL-UNNAMED --add-opens=java.base/sun.
security.action=ALL-UNNAMED --add-opens=java.ba...
```

```
#  Get current configurations via PySpark
configurations = spark.sparkContext.getConf().getAll()
for conf in configurations:
      print(conf)
```

The output you will receive could be similar to this:

```
('spark.driver.extraJavaOptions', '-XX:+IgnoreUnrecognizedVMOptions
--add-opens=java.base/java.lang=ALL-UNNAMED --add-opens=java.base/java.
lang.invoke=ALL-UNNAMED --add-opens=java.base/java.lang.reflect=ALL-
```

```
UNNAMED --add-opens=java.base/java.io=ALL-UNNAMED --add-opens=java.
base/java.net=ALL-UNNAMED --add-opens=java.base/java.nio=ALL-UNNAMED
--add-opens=java.base/java.util=ALL-UNNAMED --add-opens=java.base/java.
util.concurrent=ALL-UNNAMED --add-opens=java.base/java.util.concurrent.
atomic=ALL-UNNAMED --add-opens=java.base/sun.nio.ch=ALL-UNNAMED --add-
opens=java.base/sun.nio.cs=ALL-UNNAMED --add-opens=java.base/sun.security.
action=ALL-UNNAMED --add-opens=java.base/sun.util.calendar=ALL-UNNAMED
--add-opens=java.security.jgss/sun.security.krb5=ALL-UNNAMED')
('spark.app.submitTime', '1662916389744')
('spark.sql.warehouse.dir', 'file:/Users/.../spark-warehouse')
('spark.app.id', 'local-1662916391188')
('spark.executor.id', 'driver')
('spark.app.startTime', '1662916390211')
('spark.app.name', 'PySparkShell')
('spark.driver.port', '54543')
('spark.sql.catalogImplementation', 'hive')
('spark.rdd.compress', 'True')
('spark.executor.extraJavaOptions', '-XX:+IgnoreUnrecognizedVMOptions
--add-opens=java.base/java.lang=ALL-UNNAMED --add-opens=java.base/java.
lang.invoke=ALL-UNNAMED --add-opens=java.base/java.lang.reflect=ALL-
UNNAMED --add-opens=java.base/java.io=ALL-UNNAMED --add-opens=java.
base/java.net=ALL-UNNAMED --add-opens=java.base/java.nio=ALL-UNNAMED
--add-opens=java.base/java.util=ALL-UNNAMED --add-opens=java.base/java.
util.concurrent=ALL-UNNAMED --add-opens=java.base/java.util.concurrent.
atomic=ALL-UNNAMED --add-opens=java.base/sun.nio.ch=ALL-UNNAMED --add-
opens=java.base/sun.nio.cs=ALL-UNNAMED --add-opens=java.base/sun.security.
action=ALL-UNNAMED --add-opens=java.base/sun.util.calendar=ALL-UNNAMED
--add-opens=java.security.jgss/sun.security.krb5=ALL-UNNAMED')
('spark.serializer.objectStreamReset', '100')
('spark.driver.host', '192.168.0.16')
('spark.master', 'local[*]')
('spark.submit.pyFiles', '')
('spark.submit.deployMode', 'client')
('spark.ui.showConsoleProgress', 'true')
```

In a similar way, you can set the Spark configuration parameters during runtime:

```scala
// Set the Spark configuration parameters during runtime in Scala
spark.conf.set("spark.sql.shuffle.partitions", "30")
```

```python
# Set the Spark configuration parameters during runtime in PySpark
spark.conf.set("spark.app.name", "Hands-On Spark 3")
spark.conf.get("spark.app.name")
```

```
Output: 'Hands-On Spark 3'
```

You can also use the SparkSession to work with the catalog metadata, via the catalog variable and spark.catalog.listDatabases and spark.catalog.listTables methods:

```scala
// List Spark Catalog Databases
val ds = spark.catalog.listDatabases
ds.show(false)
```

```
+-------+----------------+------------------------------------+
|name   |description     |locationUri                         |
+-------+----------------+------------------------------------+
|default|default database|file:/Users/.../spark-warehouse     |
+-------+----------------+------------------------------------+
```

```scala
// List Tables Spark Catalog
val ds = spark.catalog.listTables
ds.show(false)
```

```
+------------+--------+-----------+---------+-----------+
|name        |database|description|tableType|isTemporary|
+------------+--------+-----------+---------+-----------+
|hive_table  |default |null       |MANAGED  |false      |
|sample_table|null    |null       |TEMPORARY|true       |
|table_1     |null    |null       |TEMPORARY|true       |
+------------+--------+-----------+---------+-----------+
```

SparkSession in spark-shell

The SparkSession object is created by the Spark driver program. Remember we mentioned in a previous section that the SparkSession object is automatically created for you when you use the Spark shell and it is available via the spark variable. You can use the spark variable in the Spark shell command line like this:

```
scala> spark.version
Spark Version : 3.3.0
```

Create a SparkSession Programmatically

The more secure way of creating a new SparkSession in Scala or PySpark is to use the object org.apache.spark.sql.SparkSession.Builder with the constructor builder() while at the same time calling the getOrCreate() method. Working this way ensures that if a SparkSession already exists, it is used; otherwise, a new one is created:

```scala
// Scala code to create a SparkSession object
import org.apache.spark.sql.SparkSession
object NewSparkSession extends App {
  val spark = SparkSession.builder()
      .master("local[4]")
      .appName("Hands-On Spark 3")
      .getOrCreate();
  println(spark)
  println("The Spark Version is : "+spark.version)
}
```

```
org.apache.spark.sql.SparkSession@ddfc241
The Spark Version is : 3.3.0
```

```python
# PySpark code to create a SparkSession object
import pyspark
from pyspark.sql import SparkSession
spark = SparkSession.builder.master("local[4]") \
                    .appName("Hands-On Spark 3") \
                    .getOrCreate()
```

```
print(spark)
print("Spark Version : "+spark.version)
Spark Version : 3.3.0
```

2.4 Transformations, Actions, Immutability, and Lazy Evaluation

The Spark core data structures, RDD (Resilient Distributed Dataset), and dataframes are *immutable* in nature; it means once they are created, they cannot be modified. In this context, *immutable* is a synonym of *unchangeable*. Spark operations in distributed datasets are classified as transformations and actions.

Transformations

Transformations are operations that take a RDD or dataframe as input and return a new RDD or dataframe as output. Therefore, transformations preserve the original copy of the data, and that is why Spark data structures are said to be immutable. Another important characteristic of the transformations is that they are not executed immediately after they are defined; on the contrary, they are memorized, creating a transformations lineage as the one shown in Figure 2-11. For example, operations such as map(), filter(), and others don't take effect until an action is defined.

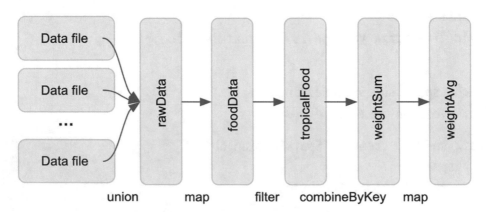

Figure 2-11. *Example of a transformations lineage*

A transformations lineage means the sequence of operations applied are recorded in a diagram called a DAG (Directed Acyclic Graph) (Figure 2-12) and executed only when an action is triggered. This idea of deferring transformations until an action takes place is what it is known as *lazy evaluation* of the transformations. Put in a simple way, when you define operations as those mentioned before, nothing happens until you instruct Spark what to do with the data. Lazy evaluation is the way Spark optimizes operations, because it allows Spark to select the best way to execute them when the complete workflow is defined.

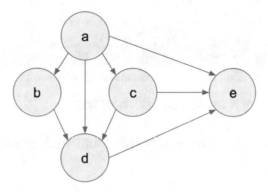

Figure 2-12. *An example of a Directed Acyclic Graph*

The following are some basic transformations in Spark:

- map()
- flatMap()
- filter()
- groupByKey()
- reduceByKey()
- sample()
- union()
- distinct()

Transformations can define narrow dependencies or wide dependencies; therefore, Spark implements two types of transformations, narrow transformations and wide transformations.

Narrow Transformations

Narrow transformations are operations without data shuffling, that is to say, there is no data movement between partitions. Thus, narrow transformations operate on data residing in the same partition as can be seen in Figure 2-13.

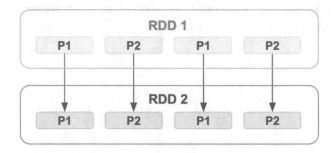

Figure 2-13. *Example of narrow transformations*

Narrow transformations are the result of functions such as map(), mapPartition(), flatMap(), filter(), or union().

Wide Transformations

Wide transformations are operations involving data shuffling; it means there is data movement between partitions as can be seen in Figure 2-14.

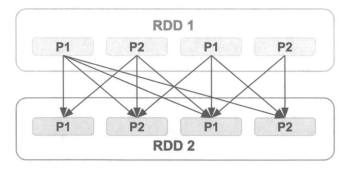

Figure 2-14. *Example of wide transformations*

Wide transformations are more costly in terms of computing and network resources because they implicate shuffle operations, meaning data must be moved across the network or at least between partitions.

Wide operations are the result of functions such as groupByKey(), join(), distinct(), aggregateByKey(), aggregate(), repartition(), or intersect().

Actions

Actions, on the other hand, are Spark operations returning a single value—in other words, operations not returning another data structure, RDD, or dataframe. When an action is called in Spark, it triggers the transformations preceding it in the DAG.

Examples of functions triggering actions are aggregate(), collect(), count(), fold(), first(), min(), max(), top(), etc.

2.5 Summary

In this chapter we have covered the more essential steps to have Spark up and running: downloading the necessary software and configuration. We have also seen how to work with the Spark shell interface and how to execute self-contained applications and examples using the Spark shell interface. Finally, we went through the Spark concepts of immutability, lazy evaluation, transformations, and actions. In the next chapter, we explain the Spark low-level API together with the notion of Resilient Distributed Datasets (RDDs).

CHAPTER 3

Spark Low-Level API

Although the term *application programming interface (API)* is mostly used regarding web services or resources shared through the Web, when it comes to Apache Spark, it possesses an additional meaning, referring to the way users can interact with the framework.

Spark has several APIs for different purposes. In this chapter we are going to study the so-called low-level API or Spark Core API, which facilitates users' direct manipulation of the Spark Resilient Distributed Datasets (RDDs), which are the Spark building blocks for the other Spark data structures of higher level such as DataFrames and datasets.

3.1 Resilient Distributed Datasets (RDDs)

The Resilient Distributed Datasets (RDDs), datasets, DataFrames, and SQL tables are the Spark core abstractions available; nevertheless, RDDs are the main Spark core abstraction. RDDs are immutable collections of objects, meaning once they are created, they cannot be changed. *Immutable* also means that any operation over an existing RDD returns a new RDD, preserving the original one.

Datasets handled as RDDs are divided into logical partitions (as seen in Figure 3-1) that can be processed in parallel across different nodes of the cluster using a low-level API. To manage parallel processing and logical partitioning, RDDs provide the concept of abstraction; thus, you do not have to worry about how to deal with them. Other Spark data entities such as dataframes and datasets are built on top of RDDs. The operations supported by Spark RDDs are transformations and actions.

Additionally, RDDs are fault-tolerant entities, meaning they possess self-recovery capacity in case of a failure. RDDs operate over fault-tolerant file systems like GFS, HDFS, AWS S3, etc. If any RDD partition breaks down, operations can continue, recovering the data from another one. On top of that, when Spark runs in a cluster like YARN, for example, it provides additional failure protection as Spark can recuperate from disasters.

© Alfonso Antolínez García 2023
A. Antolínez García, *Hands-on Guide to Apache Spark 3*, https://doi.org/10.1007/978-1-4842-9380-5_3

RDDs can be created by parallelizing already existing collections or from external datasets, such as text, sequence, CSV, and JSON files in a local file system, HDFS, AWS S3, HBase, Cassandra, or any other Hadoop-compatible input data source.

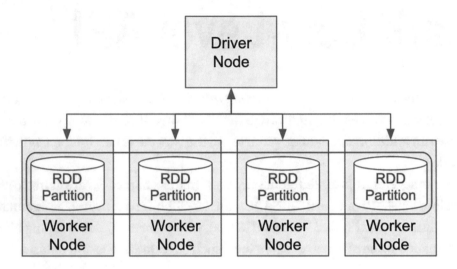

Figure 3-1. *RDD logical partitions*

We are going to explore the three methods of creating Spark RDDs.

Creating RDDs from Parallelized Collections

Parallelized collections can be created by calling the parallelize method of the SparkContext on an existing collection of the driver program. In doing this, the collection of elements are transformed into a distributed dataset and can be processed in parallel. You can use the `sparkContext.parallelize()` function to create RDDs from a collection of elements.

In Figure 3-2 it is graphically shown how the `sparkContext.parallelize()` function works when transforming a list of elements into a RDD.

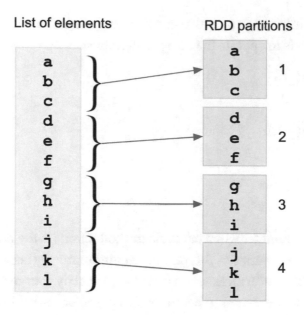

Figure 3-2. *Creating a RDD using sparkContext.parallelize()*

Next, you can see how you can implement the RDD depicted in Figure 3-2 with PySpark:

```
alphabetList = ['a','b','c','d','e','f','g','h','i','j','k','l']
rdd = spark.sparkContext.parallelize(dataList, 4)
print("Number of partitions: "+str(rdd.getNumPartitions()))
Number of partitions: 4
```

The following is another example, this time in Scala, of how to create a RDD by parallelizing a collection of numbers:

```
// A Scala example of RDD from a parallelized collection
val myCollection = Array(1, 2, 3, 4, 5, 6, 7, 8, 9, 10)
val rdd = spark.sparkContext.parallelize(myCollection)
```

Remember from the previous chapter the PySpark shell automatically provides the SparkContext "sc" variable; thus, we can use `sc.parallelize()` to create a RDD:

```
# A PySpark RDD from a list collection
rdd = sc.parallelize([1,2,3,4,5,6,7,8,9,10])
```

Now rdd is a distributed dataset that can be manipulated in parallel. Therefore, you can use RDD functions to operate the array of elements:

```
scala> rdd.reduce(_ + _)
res7: Int = 55

scala> rdd.reduce(_ min _)
res8: Int = 1

scala> rdd.reduce(_ max _)
res9: Int = 10
```

One important parameter when parallelizing collections is the number of partitions to slice the dataset into. Spark sets this parameter automatically according to the cluster available; nevertheless, you can always specify your own number of partitions, passing the number as a second parameter to `sc.parallelize()` (e.g., `sc.parallelize(myCollection, 4)`).

Creating RDDs from External Datasets

Spark can create a RDD from any Hadoop-compatible source, for example:

- Local file system

- HDFS

- Cassandra

- HBase

- AWS S3

- Etc.

Spark supports numerous file formats like

- Text files

- Sequence files

- CSV files

- JSON files

- And any Hadoop input format

A RDD can be created from a text file using the SparkContext's textFile method. This method takes as parameter the URL of the file, either its path in case of using a local file system or a URI gs://, hdfs://, s3://, etc. in case of accessing it from a distributed file system. For example, here's for a file located in your local file system:

```
scala> val readmeFile =
sc.textFile("/YOUR/SPARK/HOME/README.md")
readmeFile: org.apache.spark.rdd.RDD[String] =
/YOUR/SPARK/HOME/README.md MapPartitionsRDD[7] at textFile at <console>:26
```

If you are using a file located in your local file system, it must be available on the same path to all the nodes. Thus, you have two options: either you copy it to each worker node or use a network-shared file system such as HDFS. The following is an example of how you can load a file located in a distributed file system:

```
scala> val myFile = sc.textFile("gs://${BUCKET_NAME}/FILE.txt")
```

When working with files, it is common to distribute the information across multiple files because appending all the information to just one of them could result in a size difficult to manage, or you can be interested in splitting the information among different files, because every file could have a meaningful name and so on. This often results in folders with files that should be operated collectively to have meaningful information. To facilitate this operation, all of Spark's file-based input methods support folders, compressed files, and wildcards. For example, the method textFile() shown before can be used as textFile("/path/"), textFile("/path/*.csv"), textFile("/path/*.gz"), etc. When multiple files are used as input, the order of the partitions created by Spark depends on the order the files are uploaded from the file system.

You can also control the number of partitions a read file is divided into. For example, the textFile() method accepts a second parameter to specify the number of partitions, as you can see in the following example:

```
scala> val myFile = sc.textFile("/YOUR/SPARK/HOME/README.md", 10)
```

By default, Spark will split the file's data into chunks of the same number as file blocks, but as you have just seen, you can request Spark to divide your file into more partitions. However, what you cannot do is to request from Spark fewer partitions than file blocks.

Apart from the textFile() method, the Spark Scala API also supports other input data formats. For example, the wholeTextFiles() method can be used to read multiple small UTF-8-encoded text files from HDFS, a local file system, or any other

Hadoop-compatible URI. While `textFile()` reads one or more files and returns one record per line of each file processed, the `wholeTextFiles()` method reads the files returning them as a key-value pair (*path of the file, file content*), hence preserving the relationship between the content and the file of origin. The latter might not happen when `textFile()` processes multiple files at once, because the data is shuffled and split across several partitions. Because the process of sequentially processing files depends on the order they are returned by the file system, the distribution of rows within the file is not preserved.

Since each file is loaded in memory, `wholeTextFiles()` is preferred for small file processing. Additionally, `wholeTextFiles()` provides a second parameter to set the minimum number of partitions.

The Apache Spark API also provides methods to handle Hadoop sequence files. This Hadoop file format is intended to store serialized key-value pairs. Sequence files are broadly used in MapReduce processing tasks as input and output formats. The sequence file format offers several advantages such as compression at the level of record and block. They can be used to wrap up a large number of small files, thus solving the drawback of some file systems in processing large numbers of small files.

Apache Spark also provides a method to save RDDs as serialized Java objects, a format similar to the Hadoop sequence files mentioned just before. `RDD.saveAsObjectFile` and `SparkContext.objectFile` methods can be used to save and load RDDs. `saveAsObjectFile()` uses Java serialization to store information on a file system and permits saving metadata information about the data type when written to a file. The following is an example of how `saveAsObjectFile()` and `SparkContext.objectFile()` can be employed to save and recover a RDD object:

```scala
scala> val list = sc.parallelize(List("España","México","Colombia","Perú","
Ecuador"))
list: org.apache.spark.rdd.RDD[String] = ParallelCollectionRDD[20] at
parallelize at <console>:23

scala> list.saveAsObjectFile("/tmp/SpanishCountries")

scala> val newlist = sc.objectFile[String]("/tmp/SpanishCountries")
newlist: org.apache.spark.rdd.RDD[String] = MapPartitionsRDD[24] at
objectFile at <console>:23

scala> newlist.collect
res9: Array[String] = Array(Ecuador, España, México, Colombia, Perú)
```

Creating RDDs from Existing RDDs

Remember RDDs are immutable; hence, they cannot be changed. However, also remember we can produce new RDDs by applying transformations to the original one. A RDD can be created from another one taking advantage of transformations, for example: `map()`, `filter()`, `count()`, `distinct()`, `flatMap()`, etc.

The following is an example of how to create a new RDD from an existing one. In our first step, we create a sequence of seasons. In the second step, we create a RDD from the previous sequence using parallelize() and divide it into four partitions. In the third step, we produce the findSeasons RDD from the seasonsParallel one, by extracting the first letter of the previous elements. Finally, we show the content of the findSeasons RDD and check the number of partitions findSeasons is split into. We use the `collect()` method to first bring the RDD elements to the driver node:

```scala
scala> val seasonsCollection = Seq("Summer", "Autumn", "Spring", "Winter")
seasonsCollection: Seq[String] = List(Summer, Autumn, Spring, Winter)

scala> val seasonsParallel =spark.sparkContext.parallelize(seasons
Collection,4)
seasonsParallel: org.apache.spark.rdd.RDD[String] =
ParallelCollectionRDD[4] at parallelize at <console>:23

scala> val findSeasons= seasonsParallel.map(s -> (s.charAt(0), s))
findSeasons: org.apache.spark.rdd.RDD[(Char, String)] = MapPartitionsRDD[5]
at map at <console>:23

scala> findSeasons.collect().foreach(c => println(c))
(S,Spring)
(W,Winter)
(S,Summer)
(A,Autumn)
scala> println("Partitions: " + findSeasons.getNumPartitions)
Partitions: 4
```

The use of `collect()` is dangerous because it collects all the RDD data from all the workers in the driver node; thus, you can run out of memory if the size of the whole dataset does not fit into the driver memory. It is very inefficient as well, because all the data from the cluster has to travel through the network, and this is much slower than writing to disk and much more inefficient than computation in memory. If you only want to see some samples from your RDD, it is safer to use the `take()` method:

```scala
scala> findSeasons.take(2).foreach(c => println(c))
(S,Summer)
(A,Autumn)
```

3.2 Working with Key-Value Pairs

Some Spark RDD operations are only available for key-value pair data formats. These operations are called pair RDD operations, and for them, Spark provides various Pair RDD Functions, members of the PairRDDFunctions class, to handle RDD key-value pairs. The prototypical pair RDDs are those that imply distributed reorganization of data including Pair RDD Transformation Functions related to grouping or aggregating elements by their keys.

On the other hand, a key-value pair is a data type represented by a collection of two joined data elements: a key and a value. The key is a unique identifier of a data object. The value is a variable belonging to the dataset. An example of key-value pairs could be a telephone directory, where a person's or business's name is the key and the phone number(s) is the value. Another example could be a car's catalog in which the car could be the key and its attributes (model, color, etc.) could be the values. Key-value pairs are commonly used for log and configuration files.

Creating Pair RDDs

Pair RDDs can be created using the `map()` function that returns a key-value pair. However, the procedure can change depending on the language. In Scala, for example, to be able to take advantage of the Pair RDD Functions, you need to have your data in the form of tuples. In the following you can see a Scala example of how to get it:

```
val spark = SparkSession.builder()
   .appName("Hands-On Spark 3")
   .master("local[2]")
   .getOrCreate()

val currencyListRdd = spark.sparkContext.parallelize(List("USD;Euro;GBP;
CHF","CHF;JPY;CNY;KRW","CNY;KRW;Euro;USD","CAD;NZD;SEK;MXN"))
val currenciesRdd = currencyListRdd.flatMap(_.split(";"))
val pairRDD = currenciesRdd.map(c=>(c,1))
pairRDD.foreach(println)
(USD,1)
(Euro,1)
(GBP,1)
(CHF,1)
(CHF,1)
(JPY,1)
(CNY,1)
(KRW,1)
(CNY,1)
(KRW,1)
(Euro,1)
(USD,1)
(CAD,1)
(NZD,1)
(SEK,1)
(MXN,1)
```

The preceding code first creates a session of name "Hands-On Spark 3" using the .appName() method and a local cluster specified by the parameter local[n], where n must be greater than 0 and represents the number of cores to be allocated, hence the number of partitions, by default, RDDs are going to be split up into. If a SparkSession is available, it is returned by getOrCreate(); otherwise, a new one for our program is created.

Next, the same example is reproduced but using PySpark this time:

```
currencyList = ["USD;Euro;GBP;CHF","CHF;JPY;CNY;KRW","CNY;KRW;Euro;USD",
"CAD;NZD;SEK;MXN"]
currencyListRdd = spark.sparkContext.parallelize(currencyList, 4)

currenciesRdd = currencyListRdd.flatMap(lambda x: x.split(";"))
pairRDD = currenciesRdd.map(lambda x: (x,1))
sampleData = pairRDD.take(5)

for f in sampleData:
    print(str("("+f[0]) +","+str(f[1])+")")
```

```
(USD,1)
(Euro,1)
(GBP,1)
(CHF,1)
(CHF,1)
```

If you want to show the full list, use the collect() method instead of take() like this:

```
sampleData = pairRDD.collect()
```

But be careful. In large datasets, this could cause you overflow problems in your driver node.

Showing the Distinct Keys of a Pair RDD

You can use distinct() to see all the distinct keys in a pair RDD. First, we show a Scala code snippet revealing the distinct keys in a list of currencies:

```
pairRDD.distinct().foreach(println)
(MXN,1)
(GBP,1)
(CHF,1)
(CNY,1)
(KRW,1)
(SEK,1)
(USD,1)
```

```
(JPY,1)
(Euro,1)
(NZD,1)
(CAD,1)
```

Now, here's another code snippet in PySpark to get the same result:

```
# Returning the distinct keys.
sampleData = pairRDD.distinct().collect()
for f in sampleData:
    print(str("("+f[0]) +","+str(f[1])+")")
```

```
(GBP,1)
(MXN,1)
(CNY,1)
(KRW,1)
(USD,1)
(Euro,1)
(CHF,1)
(JPY,1)
(CAD,1)
(NZD,1)
(SEK,1)
```

As you can see in the preceding example, keys are not necessarily returned sorted. If you want to have your returned data ordered by key, you can use the sorted() method. Here is an example of how you can do it:

```
sampleData = sorted(pairRDD.distinct().collect())
for f in sampleData:
    print(str("("+f[0]) +","+str(f[1])+")")
```

```
(CAD,1)
(CHF,1)
(CNY,1)
(Euro,1)
(GBP,1)
(JPY,1)
```

```
(KRW,1)
(MXN,1)
(NZD,1)
(SEK,1)
(USD,1)
```

Transformations on Pair RDDs

In this section we are going to review several of the more important transformations that can be executed on pair RDDs.

We have already mentioned RDDs are immutable in nature; therefore, transformation operations executed on a RDD return one or several new RDDs without modifying the original one, hence creating a RDD lineage, which is use by Spark to optimize code execution and to recover from a failure. Apache Spark takes advantage of RDD lineage to rebuild RDD partitions lost. A graphical representation of a RDD lineage or RDD dependency graph can be seen in Figure 3-3.

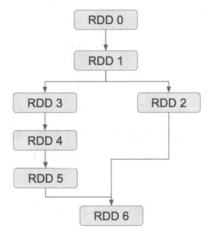

Figure 3-3. *Example of Apache Spark RDD lineage*

Sorting Pair RDDs by Key

The method sortByKey() sorts a pair RDD. In Scala, it could be written like this:

```
pairRDD.sortByKey().foreach(println)
(KRW,1)
(KRW,1)
```

```
(CNY,1)
(CNY,1)
(GBP,1)
(NZD,1)
(JPY,1)
(MXN,1)
(Euro,1)
(Euro,1)
(SEK,1)
(USD,1)
(USD,1)
(CAD,1)
(CHF,1)
(CHF,1)
```

sortByKey() admits two parameters, ascending (true/false) sorting and the numPartitions, to set the number of partitions that should be created with the results returned by sortByKey():

```
pairRDD.sortByKey(true).foreach(println)
```

In PySpark, we can use the following code snippet to achieve the same result:

```
sampleData = pairRDD.sortByKey().collect()
for f in sampleData:
    print(str("("+f[0]) +","+str(f[1])+")")
(CAD,1)
(CHF,1)
(CHF,1)
(CNY,1)
(CNY,1)
(Euro,1)
(Euro,1)
(GBP,1)
(JPY,1)
(KRW,1)
(KRW,1)
```

```
(MXN,1)
(NZD,1)
(SEK,1)
(USD,1)
(USD,1)
```

Adding Values by Key in a RDD

The Spark PairRDDFunction reduceByKey() is a wide transformation that shuffles the data of all RDD partitions. It merges the values of each key in a RDD using an associated reduction function. The reduceByKey() is optimized for large datasets, because Spark can combine the output by key before shuffling the information.

The reduceByKey() syntax is as follows:

```
sparkContext.textFile("hdfs://")
                .flatMap(line => line.split("ELEMENT_SEPARATOR"))
                .map(element => (element,1))
                .reduceByKey((a,b)=> (a+b))
```

To illustrate the power of this function, we are going to use a portion of the *Don Quixote of La Mancha* to have a larger dataset. You have already seen how to load files and transform them into RDDs. So let's start with an example in Scala:

```
val DonQuixoteRdd = spark.sparkContext.textFile("DonQuixote.txt")
DonQuixoteRdd.foreach(println)
// You would see an output like this
saddle the hack as well as handle the bill-hook. The age of this
In a village of La Mancha, the name of which I have no desire to call
gentleman of ours was bordering on fifty; he was of a hardy habit,
spare, gaunt-featured, a very early riser and a great sportsman. They
to mind, there lived not long since one of those gentlemen that keep a
will have it his surname was Quixada or Quesada (for here there is some
lance in the lance-rack, an old buckler, a lean hack, and a greyhound
difference of opinion among the authors who write on the subject),
for coursing. An olla of rather more beef than mutton, a salad on most
although from reasonable conjectures it seems plain that he was called
nights, scraps on Saturdays, lentils on Fridays, and a pigeon or so
Quexana. This, however, is of but little importance to our tale; it
```

extra on Sundays, made away with three-quarters of his income. The rest
of it went in a doublet of fine cloth and velvet breeches and shoes to
will be enough not to stray a hair's breadth from the truth in the
telling of it.
match for holidays, while on week-days he made a brave figure in his
best homespun. He had in his house a housekeeper past forty, a niece
under twenty, and a lad for the field and market-place, who used to

```scala
val wordsDonQuixoteRdd = DonQuixoteRdd.flatMap(_.split(" "))
val tupleDonQuixoteRdd = wordsDonQuixoteRdd.map(w => (w,1))
val reduceByKeyDonQuixoteRdd = tupleDonQuixoteRdd.reduceByKey((a,b)=>a+b)
// Finally, you can see the values merged by key and added.
// The output has been truncated.
reduceByKeyDonQuixoteRdd.foreach(println)
(Quesada,1)
(went,1)
(under,1)
(call,1)
(this,1)
...
(made,2)
(it,4)
(on,7)
(he,3)
(in,5)
(for,3)
(the,9)
(a,15)
(or,2)
(was,4)
(to,6)
(breeches,1)
(more,1)
(of,13)

println("Count : "+reduceByKeyDonQuixoteRdd.count())
Count : 157
```

As usual, you can achieve the same results employing PySpark code. Let me show it to you with an example. In this case most of the outputs have been suppressed, but believe me the final result is the same:

```
DonQuixoteRdd = spark.sparkContext.textFile("DonQuixote.txt")
DonQuixoteRdd2 = DonQuixoteRdd.flatMap(lambda x: x.split(" "))
DonQuixoteRdd3 = DonQuixoteRdd2.map(lambda x: (x,1))
DonQuixoteRddReduceByKey = DonQuixoteRdd3.reduceByKey(lambda x,y: x+y)
print("Count : "+str(DonQuixoteRddReduceByKey.count()))
Count : 157
```

Saving a RDD as a Text File

Though saving an existing RDD to a file is an action rather than a transformation, we are going to introduce it here, to take advantage of the DonQuixote RDD to show you how to save in-memory data to a fault-tolerant device.

You can save your RDDs as a string representation of elements using the saveAsTextFile() method. saveAsTextFile() will store the RDD as a text file.

saveAsTextFile(path: str, compressionCodecClass: Optional) can take two parameters. One of them is mandatory, *path*, which according to the official documentation represents "*path to text file*"; however, in fact it is a folder. Spark writes the RDD split into different files along with the success file (_success). The files are named part-00000, part-00001, and so on.

compressionCodecClass permits specifying a compression codec to store your data compressed.

Following with our DonQuixote example, let's write our RDD to a file:

```
reduceByKeyDonQuixoteRdd.saveAsTextFile("RDDDonQuixote")
```

You can also create a temporary directory to store your files, and instead of letting your operating system decide where to make that directory, you can have control over those parameters. Here is an example in PySpark:

```
import tempfile
from tempfile import NamedTemporaryFile

tempfile.tempdir = "./"
RDDDonQuixote = NamedTemporaryFile(delete=True)
```

```
RDDDonQuixote.close()
DonQuixoteRdd3.saveAsTextFile(RDDDonQuixote.name)
print(RDDDonQuixote)
print(RDDDonQuixote.name)
# Output
<tempfile._TemporaryFileWrapper object at 0x7f9ed1e65040>
/Users/aantolinez/tmp906w7eoy

from fileinput import input
from glob import glob
''.join(sorted(input(glob(RDDDonQuixote.name + "/part-0000*"))))
# Output
"('(for', 1)\n('An', 1)\n('Fridays,', 1)\n('He', 1)\n('I', 1)\n
('In', 1)\n('La', 1)\n('Mancha,', 1)\n('Quesada', 1)\n('Quexana.', 1)\n
('Quixada', 1)\n('Saturdays,', 1)\n('Sundays,', 1)\n('The', 1)\n
('The', 1)\n('They', 1)\n('This,', 1)\n('a', 1)\n('a', 1)\n('a', 1)\n
('a', 1)\n('a', 1)\n('a', 1)\n('a', 1)\n..."
```

In the preceding example, we have used the NamedTemporaryFile() function to create a file with a visible name in the file system. The delete parameter can take True/False values. Setting it to False, we can close the file without it being destroyed, allowing us to reopen it again later on.

Combining Data at Each Partition

One of the most common problems when working with key-value pairs is grouping and aggregating values by a standard key. In this section, we are going to use the aggregateByKey() function for aggregating data at each partition.

As with reduceByKey(), in aggregateByKey() data is combined by a common key at each partition before it is shuffled; however, reduceByKey() is kind of a particular case of aggregateByKey() in the sense that the result of the combination inside individual partitions is of the same type as the values combined. The final result, after merging the outputs of these individual combinations, is of the same type as the values of the individual combinations as well.

aggregateByKey() merges the values of a dataset by keys, and the output of this merge can be any user-defined object. With aggregateByKey() you specify how values are combined at the partition level (inside each worker) and, then, how the individual outputs from these partitions are assembled together across the nodes of the cluster to provide a final outcome.

We are going to explain this concept and the difference between reduceByKey() and aggregateByKey() with an example.

Let's assume you have the following dataset: (("a", 1), ("a", 3), ("b", 2), ("a", 5), ("b", 4), ("a", 7), ("b", 6)).

First of all we are going to create a RDD out of the preceding data:

```
val pairs = sc.parallelize(Array(("a", 1), ("a", 3), ("b", 2), ("a", 5),
("b", 4), ("a", 7), ("b", 6)))
pairs: org.apache.spark.rdd.RDD[(String, Int)] = ParallelCollectionRDD[82]
at parallelize at <console>:25
```

If you just want to add the values by key performing a sum, both reduceByKey and aggregateByKey will produce the same result. You can see an example in the following:

```
val outputReduceByKey = pairs.reduceByKey(_ + _)
outputReduceByKey.collect
outputReduceByKey: org.apache.spark.rdd.RDD[(String, Int)] =
ShuffledRDD[87] at reduceByKey at <console>:28
res49: Array[(String, Int)] = Array((a,16), (b,12))

val outputAggregateByKey = pairs.aggregateByKey(0)(_+_,_+_)
//_+_ operation inside partition, _+_ operation between partitions
outputAggregateByKey.collect
outputAggregateByKey: org.apache.spark.rdd.RDD[(String, Int)] =
ShuffledRDD[88] at aggregateByKey at <console>:27
res50: Array[(String, Int)] = Array((a,16), (b,12))
```

Let's now assume you are interested in a different sort of operation, implying the values returned are of a different kind than those of the origin. For example, imagine your desired output is a set of values, which is a different data type than the values themselves (integers) and the operations inside each partition (sum of integers returns another integer).

Next, we explain this idea with an example:

```
val outcomeSets = pairs.aggregateByKey(new HashSet[Int])(_+_, _++_)
// _+_ adds a value to a set
// _++_ joins the two sets
outcomeSets.collect
res52: Array[(String, scala.collection.mutable.HashSet[Int])] =
Array((a,Set(1, 5, 3, 7)), (b,Set(2, 6, 4)))
```

Merging Values with a Neutral ZeroValue

The foldByKey() aggregation is a kind of reduceByKey() with an initialization zero value that should not impact your final results. Like reduceByKey() it uses an associated function to combine values for each RDD's key, but additionally it gives the possibility of providing a neutral initialization value for each partition, such as 0 for addition, 1 for multiplication, or an empty list in case of concatenation of lists, that can be added to the final result an arbitrary number of times without affecting the final outcome. The zero value is initialized per key once per partition.

In the following you can see a few examples of foldByKey() usage:

```
val pairs = sc.parallelize(Array(("a", 1), ("a", 3), ("b", 2), ("a", 5),
("b", 4), ("a", 7), ("b", 6)))
pairs: org.apache.spark.rdd.RDD[(String, Int)] = ParallelCollectionRDD[82]
at parallelize at <console>:25
```

```
pairs.foldByKey(0)(_+_).collect // With init value 0
res66: Array[(String, Int)] = Array((a,24), (b,18))
```

```
pairs.foldByKey(1)(_+_).collect // With init value 1
res68: Array[(String, Int)] = Array((a,20), (b,15))
```

```
pairs.foldByKey(2)(_+_).collect // With init value 2
res66: Array[(String, Int)] = Array((a,24), (b,18))
```

Combining Elements by Key Using Custom Aggregation Functions

In this section we will explain the Spark combineByKey() generic function to combine the elements of pair RDDs by each key using custom aggregation functions. combineByKey() is a wide transformation as it requires a shuffle in the last stage.

This function turns a RDD[(K, V)] into a result of type RDD[(K, C)], for a "combined type" C, where C is the result of any aggregation of all values of key K.

In the following you can see a PySpark example of how to use `combineByKey()`:

```
pairs = sc.parallelize([("a", 1), ("a", 3), ("b", 2), ("a", 5), ("b", 4),
("a", 7), ("b", 6)])
```

```python
def to_list(x):
    return [x]
def append(x, y):
    x.append(y) # The append() method adds the y element to the x list.
    return x
def extend(x, y):
    x.extend(y) # The extend() method adds the elements of list y to the
end of the x list.
    return x
```

```
sorted(pairs.combineByKey(to_list, append, extend).collect())
```

```
[('a', [1, 3, 5, 7]), ('b', [2, 4, 6])]
```

Grouping of Data on Pair RDDs

When working with datasets of key-value pairs, a common use case is grouping all values corresponding to the same key. The `groupByKey()` method returns a grouped RDD by grouping the values by each key. The `groupByKey()` requires a function that is going to be applied to every value of the RDD.

In the following example, we convert a Scala collection type to a Spark RDD:

```scala
// Scala collection containing tuples Key-Value pairs
val countriesTuples = Seq(("España",1),("Kazakhstan",1), ("Denmark",
1),("España",1),("España",1),("Kazakhstan",1),("Kazakhstan",1))

// Converting the collection to a RDD.
val countriesDs = spark.sparkContext.parallelize(countriesTuples)

// Output
countriesTuples: Seq[(String, Int)] = List((España,1), (Kazakhstan,1),
(Denmark,1), (España,1), (España,1), (Kazakhstan,1), (Kazakhstan,1))
```

```
countriesDs: org.apache.spark.rdd.RDD[(String, Int)] =
ParallelCollectionRDD[32] at parallelize at <console>:29
```

```
countriesDs.collect.foreach(println)
// Output
(España,1)
(Kazakhstan,1)
(Denmark,1)
(España,1)
(España,1)
(Kazakhstan,1)
(Kazakhstan,1)
```

Now we will group the values by key using the groupByKey() method:

```
// Applying transformation on Pair RDD.
val groupRDDByKey = countriesDs.groupByKey()
```

```
// Output
groupRDDByKey: org.apache.spark.rdd.RDD[(String, Iterable[Int])] =
ShuffledRDD[34] at groupByKey at <console>:26
```

```
groupRDDByKey.collect.foreach(println)
```

```
// Output
(España,CompactBuffer(1, 1, 1))
(Kazakhstan,CompactBuffer(1, 1, 1))
(Denmark,CompactBuffer(1))
```

As you can see in the preceding code, groupByKey() groups the data with respect to every key, and a iterator is returned. Note that unlike reduceByKey(), the groupByKey() function doesn't perform any operation on the final output; it only groups the data and returns it in the form of an iterator. This iterator can be used to transform a key-value RDD into any kind of collection like a List or a Set.

Now imagine you want to know the number of occurrences of every country and then you want to convert the prior CompactBuffer format to a List:

```
// Occurrence of every country and transforming the CompactBuffer
to a List.
val countryCountRDD = groupRDDByKey.map(tuple => (tuple._1, tuple._2.
toList.sum))
```

```
countryCountRDD.collect.foreach(println)
```

```
// Output
(España,3)
(Kazakhstan,3)
(Denmark,1)
```

Performance Considerations of groupByKey

In some cases groupByKey() cannot be your best option to solve certain kinds of problems. For example, reduceByKey() can perform better than groupByKey() on very large datasets. Though both functions will give you the same answer, reduceByKey() is the preferred option for large datasets, because with the latter, before Spark can combine the values by key at each partition, a general shuffle is performed, which as you already know involves the movement of data across the network; hence, it is costly in terms of performance.

Let's have a look in more detail at these performance concerns with a typical word count example over a distributed dataset using reduceByKey() first and groupByKey() later on:

```
val countriesList = List("España","Kazakhstan", "Denmark","España",
"España","Kazakhstan","Kazakhstan")
```

```
val countriesDs = spark.sparkContext.parallelize(countriesList)
```

```
// Output
countriesList: List[String] = List(España, Kazakhstan, Denmark, España,
España, Kazakhstan, Kazakhstan)
countriesDs: org.apache.spark.rdd.RDD[String] = ParallelCollectionRDD[39]
at parallelize at <console>:28
```

```
val countryPairsRDD = sc.parallelize(countriesList).map(country =>
(country, 1))

val countryCountsWithReduce = countryPairsRDD
  .reduceByKey(_ + _) // reduceByKey()
  .collect()

val countryCountsWithGroup = countryPairsRDD
  .groupByKey() // groupByKey()
  .map(t => (t._1, t._2.sum))
  .collect()

// Output
countryPairsRDD: org.apache.spark.rdd.RDD[(String, Int)] =
MapPartitionsRDD[51] at map at <console>:27
countryCountsWithReduce: Array[(String, Int)] = Array((España,3),
(Kazakhstan,3), (Denmark,1))
countryCountsWithGroup: Array[(String, Int)] = Array((España,3),
(Kazakhstan,3), (Denmark,1))
```

Both functions will produce the same answer; however, reduceByKey() works better on a large dataset because Spark can combine outputs with a common key on each partition before shuffling the data across the nodes of the cluster. reduceByKey() uses a lambda function to merge values by each key on each node before the data is shuffled; after that, it merges the data at the partition level. The lambda function is used again to reduce the values returned by each partition, to obtain the final result. This behavior is showcased in Figure 3-4.

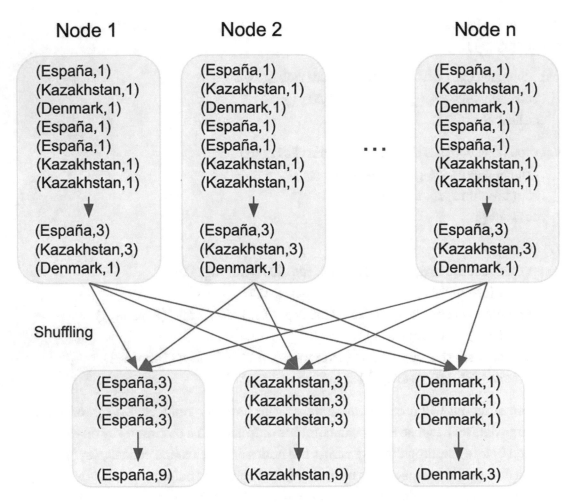

Figure 3-4. *reduceByKey internal operation*

On the other hand, when you use groupByKey(), the key-value pairs on each partition are shuffled across the nodes of the cluster. When you are working with big datasets, this behavior requires the unnecessary movement of huge amounts of data across the network representing an important process overhead.

Another reason to avoid the use of groupByKey() in large datasets are possible out-of-memory (*OutOfMemoryError*) situations in the driver node. Remember Spark must write data to disk whenever the amount of it cannot be fitted in memory. The out-of-memory situation can happen when a single executor machine receives more data that can be accommodated in its memory, causing a memory overflow. Spark saves data to disk one key at a time; thus, the process of flushing out data to a permanent storage device seriously disrupts a Spark operation.

Thus, the bigger the dataset, the more likely the occurrence of out-of-memory problems. Therefore, in general reduceByKey(), combineByKey(), foldByKey(), or others are preferable than groupByKey() for big datasets.

The groupByKey() internal operational mode is graphically shown in Figure 3-5.

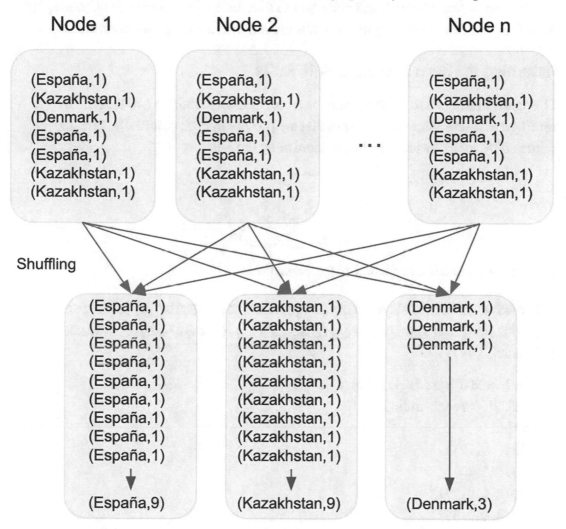

Figure 3-5. *groupByKey internal operation*

Joins on Pair RDDs

You can get the most of your key-value pair RDDs when you combine them with other key-value pair RDDs. Combining different datasets can unleash the real analytical capabilities of Apache Spark and allow you to find the insights of your data. Joining RDDs is probably one of the most typical operations you will have to perform on pair RDDs.

Returning the Keys Present in Both RDDs

The join() returns a RDD after applying a join transformation to two RDDs. The returned RDD contains only the keys that are present in both pair RDDs. The RDD returned by join() is graphically depicted in Figure 3-6.

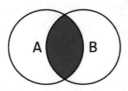

Figure 3-6. *Result set of join() transformations*

```
val rdd1 = sc.parallelize(Array(("PySpark",10),("Scala",15),("R",100)))
val rdd2 = sc.parallelize(Array(("Scala",11),("Scala",20),("PySpark",75),
("PySpark",35)))

val joinedRDD = rdd1.join(rdd2)
joinedRDD.foreach(println)
// Output
(Scala,(15,11))
(Scala,(15,20))
(PySpark,(10,75))
(PySpark,(10,35))
```

The same results can be achieved using PySpark code as you see just in the following:

```
rdd1 = spark.sparkContext.parallelize([("PySpark",10),("Scala",15),
("R",100)])
rdd2 = spark.sparkContext.parallelize([("Scala",11),("Scala",20),
("PySpark",75), ("PySpark",35)])

joinedRDD = rdd1.join(rdd2)
```

```
print(joinedRDD.collect())
# Output
[('Scala', (15, 11)), ('Scala', (15, 20)), ('PySpark', (10, 75)),
('PySpark', (10, 35))]
```

Returning the Keys Present in the Source RDD

The leftOuterJoin() returns a pair RDD having the entries of each key present in the source (left) RDD. The returned RDD has the key found in the source (left) RDD and a tuple, a combination of the value in the source RDD and one of the values of that key in the other pair RDD (right). In other words, the leftOuterJoin() returns all records from the left (A) RDD and the matched records from the right (B) RDD.

The RDD returned by leftOuterJoin() is graphically depicted in Figure 3-7.

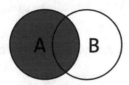

Figure 3-7. *Result set of leftOuterJoin() transformations*

In the following, you can see how to apply the leftOuterJoin() in Scala:

```
val leftJoinedRDD - rdd1.leftOuterJoin(rdd2)
leftJoinedRDD.foreach(println)
// Output
(R,(100,None))
(Scala,(15,Some(11)))
(Scala,(15,Some(20)))
(PySpark,(10,Some(75)))
(PySpark,(10,Some(35)))
```

You will obtain the same result using PySpark code as you see in the following:

```
rdd1 = spark.sparkContext.parallelize([("PySpark",10),("Scala",15),
("R",100)])
rdd2 = spark.sparkContext.parallelize([("Scala",11),("Scala",20),
("PySpark",75), ("PySpark",35)])
```

```
joinedRDD = rdd1.leftOuterJoin(rdd2)

print(joinedRDD.collect())

# Output
[('R', (100, None)), ('Scala', (15, 11)), ('Scala', (15, 20)), ('PySpark',
(10, 75)), ('PySpark', (10, 35))]
```

Returning the Keys Present in the Parameter RDD

The rightOuterJoin() is identical to the leftOuterJoin() except it returns a pair RDD
having the entries of each key present in the other (right) RDD. The returned RDD has
the key found in the other (right) RDD and a tuple, a combination of the value in the
other (right) RDD and one of the values of that key in the source (left) pair RDD. In
other words, the rightOuterJoin() returns all records from the right RDD (B) and the
matched records from the left RDD (A).

The RDD returned by rightOuterJoin() is graphically depicted in Figure 3-8.

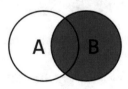

Figure 3-8. *Result set of rightOuterJoin() transformations*

Let's see how to apply rightOuterJoin() in a Scala code snippet:

```
rdd1 = spark.sparkContext.parallelize([("PySpark",10),("Scala",15),
("R",100)])
rdd2 = spark.sparkContext.parallelize([("Scala",11),("Scala",20),
("PySpark",75), ("PySpark",35)])

joinedRDD = rdd1.rightOuterJoin(rdd2)

print(joinedRDD.collect())

# Output
[('Scala', (15, 11)), ('Scala', (15, 20)), ('PySpark', (10, 75)),
('PySpark', (10, 35))]
```

Once again, you can get the same results by using PySpark code as you see in the following:

```
rdd1 = spark.sparkContext.parallelize([("PySpark",10),("Scala",15),
("R",100)])
rdd2 = spark.sparkContext.parallelize([("Scala",11),("Scala",20),
("PySpark",75), ("PySpark",35)])

joinedRDD = rdd1.leftOuterJoin(rdd2)

print(joinedRDD.collect())

# Output
[('R', (100, None)), ('Scala', (15, 11)), ('Scala', (15, 20)), ('PySpark',
(10, 75)), ('PySpark', (10, 35))]
```

Sorting Data on Pair RDDs

To sort the values of a RDD by key in ascending or descending order, you can use the Apache Spark sortByKey() transformation.

The syntax of the Spark RDD sortByKey() transformation is as follows:

RDD.sortByKey(*ascending*: Optional[bool] = True, *numPartitions*: Optional[int] = None, keyfunc: Callable[[Any], Any] = <function RDD.<lambda>>) → *pyspark.rdd. RDD[Tuple[K, V]]*

The Spark RDD sortByKey() transformation *ascending* option specifies the order of the sort (ascending order by default or when set to *true*); for descending order, you just need to set it to *false*. The *numPartitions* option specifies the number of partitions the results should be split into. The sortByKey() transformation returns a tuple of data.

Let's now see sortByKey() in action with a practical example:

```
val rdd2 = sc.parallelize(Array(("Scala",11),("Scala",20),("PySpark",75),
("PySpark",35)))

rdd1.sortByKey(true).foreach(println) // ascending order (true)
// Output
(R,100)
(PySpark,10)
(Scala,15)
```

```
rdd1.sortByKey(false).foreach(println) // descending order (false)
// Output
(PySpark,10)
(R,100)
(Scala,15)
```

Sorting a RDD by a Given Key Function

Another very important Apache Spark transformation is `sortBy()`:

```
RDD.sortBy(keyfunc: Callable[[T], S], ascending: bool = True,
numPartitions: Optional[int] = None) → RDD[T]
```

The `sortBy()` function accepts three arguments. The first one is a key function (*keyfunc*) provided, which sorts a RDD based on the key designated and returns another RDD.

The second one is a flag that specifies whether the results should be returned in ascending or descending order. The default is ascending (true).

The third parameter (*numPartitions*) specifies the total number of partitions the result is going to be divided into. *numPartitions* is an important optimization parameter, because `sortBy()` involves the shuffling of the elements of RDDs, and we have already seen it can involve unnecessary data movement.

Let's now take a look at how `sortBy()` works with an example, taking advantage of the RDD1 created from previous examples:

```
val rdd1 = sc.parallelize(Array(("PySpark",10),("Scala",15),("R",100)))
```

```
rdd1.sortBy(x => x._1).collect().foreach(println)
(PySpark,10)
(R,100)
(Scala,15)
```

```
// Output
(PySpark,10)
(R,100)
(Scala,15)
```

```
// Using now the value of the tuple to sort the data
rdd1.sortBy(x => x._2).collect().foreach(println)

// Output
(PySpark,10)
(Scala,15)
(R,100)
```

Actions on Pair RDDs

As we explained before in the book, Spark actions are RDD operations returning raw values. While transformations on a RDD return a new RDD preserving the original, actions return a value. Consequently, any operation performed in a RDD and returning anything other than a new RDD is an action.

Then, you have to also remember that RDD actions are able to trigger the effective execution of a piece of code defined in a DAG (Directed Acyclic Graph). Thus, while Spark transformations are considered lazy, meaning they are not executed right after they are defined, actions are not.

Let's have a look at some of the most important Spark actions.

Count RDD Instances by Key

The countByKey() counts the number of elements in a RDD for each key and returns a DefaultDict[key,int].

A DefaultDict is a dictionary-like object, and trying to access values that do not exist in the dictionary will return a 0 instead of throwing an error.

Let's see how it works with a Scala example. Consider the following pair RDD used in previous examples:

```
val rdd2 = sc.parallelize(Array(("Scala",11),("Scala",20),("PySpark",75),
("PySpark",35)))

rdd2.countByKey()

// Output
res48: scala.collection.Map[String,Long] = Map(PySpark -> 2, Scala -> 2)
```

```
rdd2.countByKey().foreach(println)

// Output
(PySpark,2)
(Scala,2)
```

You can access the elements of the elementsCount dictionary just as you would do for an ordinary dictionary:

```
val elementsCount = rdd2.countByKey()
println(elementsCount("Scala"))
// Output
2
```

Now you are going to see the same example, but this time using PySpark:

```
rdd2 = spark.sparkContext.parallelize([("Scala",11),("Scala",20),
("PySpark",75), ("PySpark",35)])
rdd2.collect()

# Output
[('Scala', 11), ('Scala', 20), ('PySpark', 75), ('PySpark', 35)]

# Grouping by the key, and getting the count of each group
rdd2.countByKey()

# Output
defaultdict(int, {'Scala': 2, 'PySpark': 2})

elementsCount = rdd2.countByKey()

print(elementsCount)

#Output
defaultdict(<class 'int'>, {'Scala': 2, 'PySpark': 2})
```

Now you can access the elements of the elementsCount dictionary just as you would do for an ordinary dictionary:

```
elementsCount['Scala']
# Output
2
```

Trying to access values of elementsCount that do not exist will return 0:

```
elementsCount['SQL']
# Output
0
```

Count RDD Instances by Value

The Spark countByValue() method counts each unique value in a RDD returning a dictionary of value-count pairs.

In the following, you can see how countByValue() works with a Scala example:

```
println(rdd2.countByValue())

//Output:
Map((PySpark,35) -> 1, (Scala,11) -> 1, (Scala,20) -> 1, (PySpark,75) -> 1)
```

Continuing with our previous RDD example, now we are going to see how to use countByValue() with PySpark:

```
rdd2 = spark.sparkContext.parallelize([("Scala",11),("Scala",20),
("PySpark",75), ("PySpark",35)])
sorted(rdd2.countByValue().items())

# Output
[(('PySpark', 35), 1),
 (('PySpark', 75), 1),
 (('Scala', 11), 1),
 (('Scala', 20), 1)]
```

Returning Key-Value Pairs as a Dictionary

The RDD's collectAsMap() method collects all the elements of a pair RDD in the driver node and returns key-value pairs in the RDD as a dictionary.

We are going to see the use of this method in a Scala code snippet:

```
val rdd1 = sc.parallelize(Array(("PySpark",10),("Scala",15),("R",100)))
val rdd1 = sc.parallelize(Array(("PySpark",10),("Scala",15),("R",100)))

rdd1.collectAsMap()
```

```
// Output
res62: scala.collection.Map[String,Int] = Map(R -> 100, Scala -> 15,
PySpark -> 10)
```

However, if you have duplicate keys, the last key-value pair will overwrite the former ones. In the following, the tuple ("Scala",11) has been overwritten by ("Scala",20):

```
rdd2.collectAsMap()
```

```
// Output
res63: scala.collection.Map[String,Int] = Map(Scala -> 20, PySpark -> 35)
```

Here is now the same example, but with PySpark this time:

```
rdd1 = spark.sparkContext.parallelize([("PySpark",10),("Scala",15),
("R",100)])
rdd2 = spark.sparkContext.parallelize([("Scala",11),("Scala",20),
("PySpark",75), ("PySpark",35)])
```

```
rdd1.collectAsMap()
# Output
{'PySpark': 10, 'Scala': 15, 'R': 100}
```

Remember that if you have duplicate keys, the last key-value pair will overwrite the previous ones. In the following, the tuple ("Scala",11) has been overwritten by ("Scala",20):

```
rdd2.collectAsMap()
```

```
# Output
{'Scala': 20, 'PySpark': 35}
```

Collecting All Values Associated With a Key

The Apache Spark lookup(key) method is an action that returns all values associated with a provided key in a list. It takes a key's name as a parameter:

```
val rdd2 = sc.parallelize(Array(("Scala",11),("Scala",20),("PySpark",75),
("PySpark",35)))
```

```
rdd2.lookup("PySpark")
```

```
// Output
res66: Seq[Int] = WrappedArray(75, 35)
```

Another code snippet in PySpark shows the same result:

```
rdd2 = spark.sparkContext.parallelize([("Scala",11),("Scala",20),
("PySpark",75), ("PySpark",35)])
```

```
rdd2.lookup("PySpark")
```

```
# Output
[75, 35]
```

3.3 Spark Shared Variables: Broadcasts and Accumulators

The so-called shared variables are important Spark abstractions. In simple words, shared variables are variables you can use to exchange information throughout the workers of your cluster or between your driver and the workers. In other words, these are variables intended to share information throughout the cluster.

The big data problems require the use of distributed systems. One example of these distributed infrastructures is an Apache Spark cluster, in which the driver node and executor nodes, usually, run in separate and sometimes remote computers.

In distributed computation, you are very often going to face the problem of sharing and synchronizing information across the nodes of your cluster. For instance, when you apply a function to your dataset, this function with its variables is going to be copied to every executor. As the computation in the executors runs in an independent way, the driver has no information about the update of the data contained in those variables; hence, the driver cannot track the evolution of variables copied to remote nodes.

To get around this issue, Spark provides the broadcast and accumulator variables, as a way to distribute information between executors and between executors and the driver node, allowing the driver to keep in sync with the evolution of the values contained in some variables of interest. Accumulators are used for writing data, and broadcast variables are used for reading it.

Broadcast Variables

Broadcast variables are read-only variables that allow maintaining a cached variable in each cluster node instead of transporting it with each task every time tasks are sent to the executors. Therefore, each executor will keep a local copy of the broadcast variable; in consequence, no network I/O is needed.

Broadcast variables are transferred once from the driver to the executors and used by tasks running there as many times as necessary, minimizing data movement through the network as a result because that information is not transferred to the executors every time a new task is delivered to them.

In Figure 3-9 we explain graphically the difference between using broadcast variables and normal variables to share information with the workers.

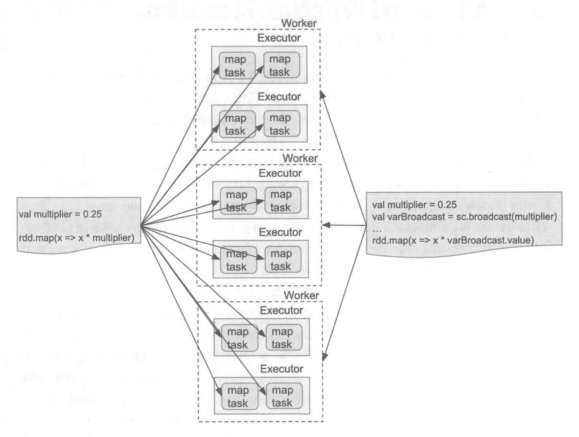

Figure 3-9. *Difference between broadcast variables and normal variables*

If you look at the left side of Figure 3-9, we use a map operation to multiply every RDD element by the external multiplier variable. In operating like this, a copy of the multiplier variable will be distributed with every task to each executor of the cluster.

On the other hand, if you look at the right side of Figure 3-9, a single copy of the broadcast variable is transmitted to each node and shared among all the tasks running on them, therefore potentially saving an important amount of memory and reducing network traffic.

Broadcast variables have the `value()` method, to store the data and access the broadcasted information.

When to Use Broadcast Variables

In the preceding section, we have used a simple example with a variable containing an integer value. However, imagine a scenario in which the external variable would constitute millions of elements. Imagine, as well, several tasks running in the same executors will use the same variable. This scenario implies copying the variable data together with the task to be executred to every executor. This operation will consume a good portion of the memory available and produce a significant network traffic surplus.

Let's now imagine a scenario like the one in section "Adding Values by Key in a RDD" working with our DonQuixote text RDD. Visualize a more exaggerated use case in which our map function launches several tasks in each executor and all of them use the external variable. In that case, several copies of the same variable would be sent to each executor.

In these circumstances, the DonQuixoteRdd text file would have to be copied to all the cluster nodes with the associated tasks. In the code snippet shown in that section, we sent the whole DonQuixoteRdd text file as a value to our functions. Therefore, working in Spark cluster execution mode, passing the whole text as a parameter represents an important network overload as it must be copied to every executor node. One of the advantages of using broadcast variables is that the data broadcasted by Spark is cached in serialized form and deserialized before running the task in each executor.

However, the use of broadcast variables only makes sense when tasks distributed across the cluster nodes need the same set of data or when caching data in deserialized format is necessary. In situations as the one depicted before, broadcast variables will reduce the volume of serialized tasks and the network traffic overhead needed to run jobs in a cluster.

One limitation of broadcast variables is that when data is broadcasted across the cluster nodes, it should not be modified if we want to be sure each executor has the exact same copy of the data.

How to Create a Broadcast Variable

A broadcast variable can be created using SparkContext's broadcast method. Let's see it with an example:

```scala
// Scala code for broadcast variables
val bVariable = sc.broadcast(Array(1, 2, 3, 4, 5, 6, 7, 8, 9))
bVariable: org.apache.spark.broadcast.Broadcast[Array[Int]] =
Broadcast(137)

bVariable.value
res70: Array[Int] = Array(1, 2, 3, 4, 5, 6, 7, 8, 9)
```

```python
# Python code for broadcast variables
bVariable =  spark.sparkContext.broadcast([1, 2, 3, 4, 5, 6, 7, 8, 9])
bVariable.value
# Output
[1, 2, 3, 4, 5, 6, 7, 8, 9]
```

Accumulators

Accumulators are variables used to track and update information across a cluster's executors. Accumulators can be used to implement counters and sums, and they can only be "added" to through associative and commutative operations.

You can see in Figure 3-10 a graphical representation of the process by which accumulators are used to collect data at the executor level and bring it to the driver node.

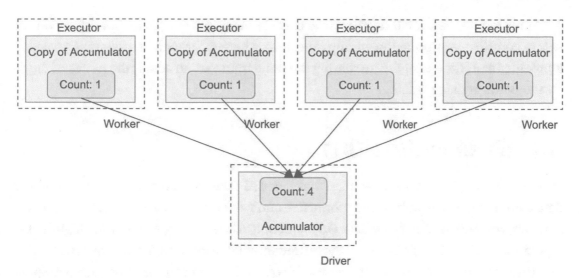

Figure 3-10. *Accumulator operation*

One important characteristic of accumulators is that executors cannot read the accumulators' value; they can only update it. The accumulator value can only be read in the driver process.

Next is a code snippet with Scala:

```
// Accumulator pyspark code snipped
val rdd = spark.sparkContext.parallelize(Array(1, 2, 3, 4, 5))
val acc = spark.sparkContext.longAccumulator("AddAccumulator")
rdd.foreach(x => acc.add(x))
print("Acc value: ", acc.value) //Value collected at the driver

// Output
(Acc value: ,15)
```

Next is the same code snippet, but this time written with PySpark:

```
# Accumulator pyspark code snipped
acc = spark.sparkContext.accumulator(0)
rdd = spark.sparkContext.parallelize([1,2,3,4,5])
rdd.foreach(lambda x:acc.add(x))
print("Acc value: ", acc.value) # Value collected at the driver

# Output
Acc value:  15
```

Summarizing, we could say that via accumulator variables, Apache Spark provides a way to coordinate information among executors, whereas through broadcast variables, it provides a method to optimize sharing the same information across the nodes of the cluster without data shuffling.

3.4 When to Use RDDs

Though it is considered that in general high-level API data structures like dataframes and datasets will allow you to be more productive and work quicker, there are circumstances in which you will need to use RDDs. On top of that, as we already said before, high-level structures are in fact built on top of these fundamental primitives. Thus, it is important to understand how they work because when you work with dataframes or datasets, Spark turns them into RDDs under the hood. Therefore, as your code becomes more complex, it is important to know the nuances of Spark RDD programming to be able to get the most out of it.

Additionally, there are some scenarios in which you will be using RDDs:

- When you need a low-level control of the physical distribution of the data partitions across the cluster

- When you need low-level transformations and actions

- When you need to make some custom operation not available in high-level APIs

- When accessing data attributes by name or column is no longer needed and hence imposing schema to your data is not strictly necessary

- When you need to maintain legacy code written using RDDs

Wrapping up, we could say that RDDs give you a more fine-grained control of your code.

3.5 Summary

In this chapter we briefly looked at the concept of the Spark low-level API, the notion of Spark Resilient Distributed Datasets (RDDs) as Spark building blocks to construct other Spark data structures such as DataFrames and datasets with a higher level of technical isolation. We also covered the most essential operations that you can perform using RDDs, and finally we also explained the so-called Spark shared variables: broadcasts and accumulators. In the next chapter, we are going to focus on the Spark high-level API and how to use it in the big data world.

CHAPTER 4

The Spark High-Level APIs

Spark SQL, dataframes, and datasets are Spark high-level API components intended for structured data manipulation, allowing Spark to automatically improve storage and computation performance. Structured data is information organized into standardized structure of schema, which makes it accessible and analyzable without further treatment. Examples of structured data are database tables, Excel sheets, RDBMS tables, Parquet files, and so on.

Spark's high-level APIs allow the optimization of applications working with a certain kind of data, like binary format files, beyond the limits permitted by Spark's RDD, for example. Dataframes and datasets take advantage of the Spark SQL's Catalyst Optimizer and Spark Project Tungsten, studied later in this chapter, to optimize their performance.

The most important difference between the Dataset API and DataFrame API is probably that the Dataset implements type safety at compile time. Datasets enact compile-time type safety, whereas DataFrames do not. Spark verifies DataFrame data types comply with those defined in its schema at runtime, whereas dataset data types are validated at compile time. We will cover the concept of compile-time type safety in detail later on in this chapter.

4.1 Spark Dataframes

Introduced in Spark 1.3, seeking improvement in the performance and scalability of Spark. The DataFrame API introduced the notion of schema to describe a data structure, allowing Spark to optimize shuffle operations by moving data across the nodes in a more efficient way. From a visual point of view, dataframes resemble relational database tables or spreadsheets as the example you can see in the following:

© Alfonso Antolínez García 2023
A. Antolínez García, *Hands-on Guide to Apache Spark 3*, https://doi.org/10.1007/978-1-4842-9380-5_4

```
+--------+----------------+-------------------+----+----+
|  Nombre|  Primer_Apellido|   Segundo_Apellido|Edad|Sexo|
+--------+----------------+-------------------+----+----+
|  Miguel|     de Cervantes|           Saavedra|  50|   M|
|Fancisco|          Quevedo|Santibáñez Villegas|  55|   M|
|    Luis|       de Góngora|          y Argote|  65|   M|
|  Teresa|Sánchez de Cepeda|         y Ahumada|  70|   F|
+--------+----------------+-------------------+----+----+
```

From a technical point of view, however, a DataFrame is a sort of view of an untyped dataset. In other words, a DataFrame is a dataset organized into columns with a header name. In Scala and Java, a DataFrame could be considered an untyped dataset of type Row (Dataset[Row]), where a Row stands for an untyped JVM object. They are a collection of rows of data organized in named columns of different data types and formed into a schema as the one you see next:

```
root
 |-- Nombre: string (nullable = true)
 |-- Primer_Apellido: string (nullable = true)
 |-- Segundo_Apellido: string (nullable = true)
 |-- Edad: integer (nullable = true)
 |-- Sexo: string (nullable = true)
```

As you can see in the preceding schema, every dataframe column includes a set of attributes such as name, data type, and a nullable flag, which represents whether it accepts null values or not.

The Dataframe API is a component of the Spark SQL module and is available for all programming languages such as Java, Python, SparkR, and Scala. Unlike RDDs, dataframes provide automatic optimization, but unlike the former, they do not provide compile-time type safety. This means that while with RDDs and datasets the compiler knows the columns' data types (string, integer, StructType, etc.), when you work with dataframes, values returned by actions are an array of rows without a defined data type. You can cast the values returned to a specific type employing Scala´s asInstanceOf() or PySpark's cast() method, for example.

Let's analyze how the implementation of type safety influences Spark application behavior with three practical examples.

For that purpose we are going to use a small dataset populated with just three of the most prominent Spanish writers of all times. First of all we are going to show how type safety influences the use of a lambda expression in a filter or map function. The following is the code snippet.

First of all we create a case class SpanishWritersDataFrame APISpanishWriters including four personal writer's attributes:

```
//spark.sparkContext.implicits._ grants access to toDF() method
import spark.sqlContext.implicits._
case class SpanishWriters(Nombre: String, Apellido: String, Edad: Int,
Sexo:String)
```

For this example we create a small dataset of Spanish writers:

```
val SpanishWritersData = Seq(SpanishWriters("Miguel", "Cervantes",
50, "M"), SpanishWriters("Fancisco", "Quevedo", 55, "M"),
SpanishWriters("Luis", "Góngora", 65, "M"))
```

In the next step, we create a RDD from the preceding set of data:

```
val SpanishWritersRDD = spark.sparkContext.parallelize(SpanishWritersData)
```

Now we use toDF() and toDS() to create a dataframe and a dataset, respectively:

```
val writersDF = SpanishWritersRDD.toDF()
val writersDS = SpanishWritersRDD.toDS()
```

Now we are going to see the differences between the data entities when using a lambda function to filter the data:

```
// Dataframe
val writersDFResult = writersDF.filter(writer => writer.Edad > 53)
// Output
error: value Edad is not a member of org.apache.spark.sql.Row val
writersDFResult = writersDF.filter(writer => writer.Edad > 53)
                                                   ^

//Dataset
val writersDSResult = writersDS.filter(writer => writer.Edad > 53)
// Output
writersDSResult: org.apache.spark.sql.Dataset[SpanishWriters] = [Nombre:
string, Apellido: string ... 2 more fields]
```

Please, pay attention to the different output we get when filtering the information in both data structures. When we apply filter to a dataframe, the lambda function implemented is returning a Row-type object and not an integer value as you probably were expecting, so it cannot be used to compare it with an integer (53 in this case). Thus, using just the column name, we cannot retrieve the value coded as a Row object. To get the Row object value, you have to typecast the value returned to an integer. Therefore, we need to change the code as follows:

```
val writersDFResult = writersDF2.filter(writer => writer.getAs[Int]
("Edad") > 53)
writersDFResult.show()
// Output
+--------+--------+----+----+
|  Nombre|Apellido|Edad|Sexo|
+--------+--------+----+----+
|Fancisco| Quevedo|  55|   M|
|    Luis| Góngora|  65|   M|
+--------+--------+----+----+
```

The preceding example shows one of the reasons datasets were introduced. The developer does not need to know the data type returned beforehand.

Another example of compile-time type safety appears when we query a nonexisting column:

```
// Dataframe
val writersDFBirthday = writersDF.select("Birthday")
// Output
rg.apache.spark.sql.AnalysisException: Column 'Birthday' does not exist.
Did you mean one of the following? [Edad, Apellido, Nombre, Sexo];

// Dataset
val writersDSBirthday = writersDS.map(writer => writer.Birthday)
// Output
error: value Birthday is not a member of SpanishWriters
val writersDSBirthday = writersDS.map(writer => writer.Birthday)
                                                       ^
```

In the preceding example, you can see the difference between execution time (dataframe) and compile time (dataset). The former will throw an error only at runtime, while the latter will give you an error message at compile time.

Another case in which we are going to find a different behavior between DataFrames and datasets is when we want to revert them to a primitive RDD. In this case DataFrame reversion to RDD won't preserve the data schema, while dataset reversion will. Let's see it again with an example:

```
// Dataframe reversion to RDD
val rddFromDF = writersDF.rdd
// Output
rddFromDF: org.apache.spark.rdd.RDD[org.apache.spark.sql.Row] =
MapPartitionsRDD[249] at rdd at <console>:75
```

However, we won't be able to work normally with this reverted RDD, because in fact the revision returns a Row of RDD. Let's use a simple operation like to see the outcome:

```
rddFromDF.map(writer => writer.Nombre).foreach(println)
// Output
error: value Nombre is not a member of org.apache.spark.sql.Row
```

Now, we are going to do the same operation, but this time with our dataset:

```
// Dataset reversion to RDD
val rddFromDS = writersDS.rdd
// Output
rddFromDS: org.apache.spark.rdd.RDD[SpanishWriters] = MapPartitionsRDD[252]
at rdd at
```

The revision returns a real RDD, so we can normally use it:

```
rddFromDS.map(writer => writer.Nombre).foreach(println)
// Output
Luis
Francisco
Miguel
```

It proves datasets preserve the data schema when reverted to RDD.

Attributes of Spark DataFrames

Like other Apache Spark modules, DataFrames were created from inception to deal with big data projects as efficiently as possible. For that reason, Spark DataFrames support being distributed across the nodes of a cluster, taking full advantage of the Spark distributed computing architecture. User SQL queries and commands sent to the DataFrames are managed by the Catalyst Optimizer, which is responsible for finding and building the query execution plan that achieves the requested result more efficiently.

Spark DataFrames incorporate many important features. One of them is the possibility to create dataframes from external sources—circumstances that are very helpful in real life, when most of the time data is going to be given in the form of files, databases, etc. Examples of the external file formats supported out of the box by Spark to load data into DataFrames can be seen in Figure 4-1.

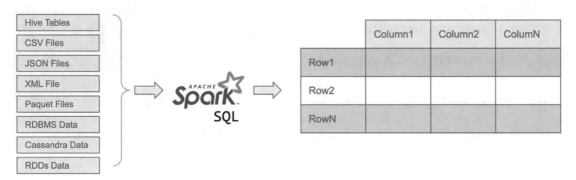

Figure 4-1. *Some out-of-the-box Spark-supported formats to load data into DataFrames*

Another important feature of DataFrames is their capacity to tackle huge volumes of data, from megabytes to petabytes. Thus, Spark DataFrames allow data management at scale.

Methods for Creating Spark DataFrames

DataFrames can be built in very different ways, such as manually, from external relational databases (MySQL, PostgreSQL, Oracle, etc.) or structured data files (CSV, Excel spreadsheets, Parquet, JSON, among others), from NoSQL databases (Hive tables, Cassandra, HBase, or MongoDB), or from already existing RDDs and binary files.

Spark provides two methods to build DataFrames manually, `toDF()` and `createDataFrame()`. Taking advantage of these methods, you can create new DataFrames from other already existing DataFrames, RDDs, datasets, lists, and sequences.

Although both methods are pretty much equivalent, they have some important differences.

Manually Creating a Spark DataFrame Using toDF()

To use `toDF()`, we first have to import Spark's `sqlContext.implicits._` library to have an implicit method to convert a RDD to a DataFrame. Let's see how to transform a RDD into a DataFrame using `toDF()`:

```
val carsData=Seq(("USA","Chrysler","Chrysler 300",292),("Germany","BMW",
"BMW 8 Series",617),("Spain", "Spania GTA", "GTA Spano",925))
val carsRdd = spark.sparkContext.parallelize(carsData) // Seq to RDD
val dfCars = carsRdd.toDF() // RDD to DF
dfCars.show()
// Output
+-------+----------+------------+---+
|     _1|        _2|          _3| _4|
+-------+----------+------------+---+
|    USA|  Chrysler|Chrysler 300|292|
|Germany|       BMW|BMW 8 Series|617|
|  Spain|Spania GTA|   GTA Spano|925|
+-------+----------+------------+---+
```

By default, `toDF()` assigns sequences "_1", "_2", "_3", "_4", and so on as column names and tries to infer data types (string and int) and flags every column as nullable, except for the numeric column. You can see this behavior by printing the dfCars dataframe schema:

```
dfCars.printSchema()
// Output
root
 |-- _1: string (nullable = true)
 |-- _2: string (nullable = true)
 |-- _3: string (nullable = true)
 |-- _4: integer (nullable = false)
```

The method toDF() accepts an indefinite number of parameters to be used as column names: df.toDF('col1', 'col2', ..., 'colN'), as you can see in the following:

```
val dfBrandedCars = carsRdd.toDF("Country","Manufacturer","Model","Power" )
dfBrandedCars.show()
// Output
+-------+------------+------------+-----+
|Country|Manufacturer|       Model|Power|
+-------+------------+------------+-----+
|    USA|    Chrysler|Chrysler 300|  292|
|Germany|         BMW|BMW 8 Series|  617|
|  Spain|  Spania GTA|   GTA Spano|  925|
+-------+------------+------------+-----+
```

The conclusion we obtain from the preceding example is that using toDF() we have no control over the dataframe schema. This means we have no control over column types and nullable flags.

Manually Creating a Spark DataFrame Using createDataFrame()

We can take advantage of the createDataFrame() method to construct DataFrames in two forms. The first one is coupling it with toDF() while taking a RDD as a parameter. Let's show how it works with an example:

```
var df2 = spark.createDataFrame(carsData) \
.toDF("Country","Manufacturer","Model","Power")
df2.show()
// Output
+-------+------------+------------+-----+
|Country|Manufacturer|       Model|Power|
+-------+------------+------------+-----+
|    USA|    Chrysler|Chrysler 300|  292|
|Germany|         BMW|BMW 8 Series|  617|
|  Spain|  Spania GTA|   GTA Spano|  925|
+-------+------------+------------+-----+
```

The second way is we can use `createDataFrame()` to create a dataframe and unleash its real power, allowing us to fully customize our own DataFrame schema. You can have a good grasp of how `createDataFrame()` works by incorporating a schema definition in the following example:

```
import org.apache.spark.sql.Row
import org.apache.spark.sql.types.{IntegerType,StringType, StructField,
StructType}
// First of all we create a schema for the carsData dataset.
val carSchema = StructType( Array(
      StructField("Country", StringType,true),
      StructField("Manufacturer", StringType,true),
      StructField("Model", StringType,true),
      StructField("Power", IntegerType,true)
))
// Notice we are using here the carsRdd RDD shown in the previous example
val carsRowRdd = carsRdd.map(carSpecs => Row(carSpecs._1, carSpecs._2,
carSpecs._3, carSpecs._4))
val dfCarsFromRDD = spark.createDataFrame(carsRowRdd,carSchema)
dfCarsFromRDD.show()
// Output
+-------+------------+------------+-----+
|Country|Manufacturer|       Model|Power|
+-------+------------+------------+-----+
|    USA|    Chrysler|Chrysler 300|  292|
|Germany|         BMW|BMW 8 Series|  617|
|  Spain|  Spania GTA|   GTA Spano|  925|
+-------+------------+------------+-----+
```

Wrapping up, we could say that though both `toDF()` and `createDataFrame()` methods can be used to create DataFrames, the former infers the data schema, while the latter gives you full customization control over the DataFrame schema.

Data Sources for Creating Spark DataFrames

Spark SQL, through the DataFrame interface, supports a wide range of data sources. A DataFrame can be operated in two different ways. The first one is directly, using relational transformations we have already seen in previous chapters. The second one is by creating a temporary view from the dataframe. The second method allows you to run SQL queries over the data, as if you were querying traditional RDBMS.

Parquet is the default data source Spark expects to use in input/output operations. This default format can be set using the `spark.sql.sources.default` property name.

Let's see it with an example. Try to load a CSV file using the `load()` method:

```
val personasDF = spark.read.load("/Users/aantolinez/Downloads/
personas.csv")
```

You will receive the following error message:

```
Caused by: java.lang.RuntimeException: file:/Users/aantolinez/Downloads/
personas.csv is not a Parquet file. Expected magic number at tail, but
found [48, 44, 70, 10]
```

However, if you load a Parquet file

```
val personasDF = spark.read.load("/Users/aantolinez/Downloads/personas.
parquet")
```

everything goes well:

```
personasDF: org.apache.spark.sql.DataFrame = [Nombre: string, Primer_
Apellido: string ... 3 more fields]
```

Exactly the same would happen if you use PySpark code:

```
personasDF = spark.read.load("/Users/aantolinez/Downloads/personas.csv")
```

```
// Output
Caused by: java.lang.RuntimeException: file:/Users/aantolinez/Downloads/
personas.csv is not a Parquet file. Expected magic number at tail, but
found [48, 44, 70, 10]
```

While using a Parquet format file, everything is fine:

```
personasDF = spark.read.load("/Users/aantolinez/Downloads/personas.
parquet")
personasDF.show(1)
# Output
+------+---------------+----------------+----+----+
|Nombre|Primer_Apellido|Segundo_Apellido|Edad|Sexo|
+------+---------------+----------------+----+----+
|Miguel|   de Cervantes|        Saavedra|  50|   M|
+------+---------------+----------------+----+----+
```

Querying Files Using SQL

Sometimes directly querying a data file instead of loading it first into a DataFrame could be interesting. Spark allows you to do it in the following way, using the same code in Scala and PySpark and getting the same result:

```
spark.sql("SELECT * FROM parquet.`/Users/aantolinez/Downloads/personas.
parquet`").show()
// Output Scala and PySpark
+--------+----------------+-------------------+----+----+
|  Nombre|  Primer_Apellido|   Segundo_Apellido|Edad|Sexo|
+--------+----------------+-------------------+----+----+
|  Miguel|     de Cervantes|           Saavedra|  50|   M|
|Fancisco|          Quevedo|Santibáñez Villegas|  55|   M|
|    Luis|       de Góngora|           y Argote|  65|   M|
|  Teresa|Sánchez de Cepeda|          y Ahumada|  70|   F|
+--------+----------------+-------------------+----+----+
```

Ignoring Corrupt and Missing Files

Spark provides the both spark.sql.files.ignoreCorruptFiles method to ignore corrupt files and spark.sql.files.ignoreMissingFiles method to ignore missing files while reading files from the file system. With the former, when set to true, Spark jobs will not crash when they find corrupted files, and the content that could have been read will still be returned. The latter means that Spark jobs will not fail when files are missing, and as in the previous method, data that could have been read will still be returned. In this context, a missing file is one that has been deleted after a DataFrame transformation has been applied.

We are going to see this Spark feature with an example:

```
// enable ignore corrupt files
spark.sql("set spark.sql.files.ignoreCorruptFiles=true")

// personas_corrupt.parquet is not real parquet file
val corruptFiles = spark.read.parquet(
      "/Users/aantolinez/Downloads/personas.parquet",
      "/Users/aantolinez/Downloads/personas_corrupt.parquet")
corruptFiles.show()
```

Time-Based Paths

Spark provides modifiedBefore and modifiedAfter options for time control over files that should be loaded at query time.

modifiedBefore takes a timestamp as a parameter instructing Spark to only read files whose modification time occurred before the given time. Similarly, modifiedAfter also takes a timestamp as a parameter but this time commanding Spark to only load files whose modification time took place after the given time. In both cases timestamp must have the following format: YYYY-MM-DDTHH:mm:ss (e.g. 2022-10-29T20:30:50).

Let's see this Spark behavior with an example in Scala and later on in PySpark:

```
val modifiedAfterDF = spark.read.format("csv")
  .option("header", "true")
  .option("modifiedAfter", "2022-10-30T05:30:00")
  .load("/Users/aantolinez/Downloads/Hands-On-Spark3");

modifiedAfterDF.show();
```

We can get the same result using PySpark code as you can see in the following:

```
modifiedAfterDF = spark.read.format("csv") \
  .option("header", "true") \
  .option("modifiedAfter", "2022-10-30T05:30:00") \
  .load("/Users/aantolinez/Downloads/Hands-On-Spark3");

modifiedAfterDF.show();
```

The output in both cases will be the same:

```
+--------+----------------+-------------------+----+----+
|  Nombre|  Primer_Apellido|  Segundo_Apellido|Edad|Sexo|
+--------+----------------+-------------------+----+----+
|  Miguel|     de Cervantes|           Saavedra|  50|   M|
|Fancisco|          Quevedo|Santibáñez Villegas|  55|   M|
|    Luis|       de Góngora|           y Argote|  65|   M|
|  Teresa|Sánchez de Cepeda|          y Ahumada|  70|   F|
+--------+----------------+-------------------+----+----+
```

Both options support the specification of a timezone via `spark.sql.session.timeZone`; in this case, timestamps will reference the timezone given.

Specifying Save Options

We have mentioned before the Spark default data source is in Parquet format; however, Spark permits interaction with many other sources of information such as JSON, ORC, CSV, and text files as well as Hive tables, Cassandra, etc. and JDBC data origins.

This large ecosystem of data origins and the Spark capacity to transform the data between different formats permit Spark to be used as an efficient ETL[1] tool. Spark can load data from the sources mentioned, transform it, and save it in the formats and repositories specified. There are four saving modes as shown in Table 4-1.

Table 4-1. *Saving Modes*

In Scala and Java	In Any Language	Meaning
SaveMode.ErrorIfExists (default)	"error" or "errorifexists" (default)	An exception is sent if data already exists at the destination when saving the DataFrame.
SaveMode.Append	"append"	Data is appended to the destination data/table.
SaveMode.Overwrite	"overwrite"	If data/table already exists in the destination, it is overwritten.
SaveMode.Ignore	"ignore"	It works similarly to the SQL CREATE TABLE IF NOT EXISTS. If data already exists in the destination, the operation is overlooked.

---

[1] ETL stands for Extract, Transform, and Load data.

Let's see how Spark saving modes work with an example:

```
val SpanishDf = spark.read.option("header", "true")
      .option("inferSchema", "true")
      .csv("/Users/aantolinez/Downloads/personas.csv")
// Writing the first DataFrame
SpanishDf.write.format("csv").mode("overwrite")
      .option("header", true)
      .save("/Users/aantolinez/Downloads/personas2.csv")
// Adding some data to append to the previous saved DataFrame
val SpanishWritersData2 = Seq(("Miguel", "de Unamuno", 70, "M"))
val SpanishWritersRdd = spark.sparkContext.parallelize(SpanishWritersData2)
val SpanishWritersAppendDF = SpanishWritersRdd.toDF()
// Appending the new data to the previous saved one
SpanishWritersAppendDF.write.format("csv").mode("append").save("/Users/
aantolinez/Downloads/personas2.csv")
```

Now if you have a look at the saved data in Figure 4-2, you see something surprising.

```
personas2.csv:
total 48
drwxr-xr-x  10 aantolinez  staff    320 Oct 30 22:01 .
drwx------@ 76 aantolinez  staff   2432 Oct 30 22:00 ..
-rw-r--r--   1 aantolinez  staff      8 Oct 30 22:01 ._SUCCESS.crc
-rw-r--r--   1 aantolinez  staff     12 Oct 30 22:00 .part-00000-7f7ac501-7011-4bf7-a684-ef347fc29f21-c000.csv.crc
-rw-r--r--   1 aantolinez  staff      8 Oct 30 22:01 .part-00000-cff368dd-78bc-4473-b70d-4037f66bedcd-c000.csv.crc
-rw-r--r--   1 aantolinez  staff     12 Oct 30 22:01 .part-00009-cff368dd-78bc-4473-b70d-4037f66bedcd-c000.csv.crc
-rw-r--r--   1 aantolinez  staff      0 Oct 30 22:01 _SUCCESS
-rw-r--r--   1 aantolinez  staff    200 Oct 30 22:00 part-00000-7f7ac501-7011-4bf7-a684-ef347fc29f21-c000.csv
-rw-r--r--   1 aantolinez  staff      0 Oct 30 22:01 part-00000-cff368dd-78bc-4473-b70d-4037f66bedcd-c000.csv
-rw-r--r--   1 aantolinez  staff     23 Oct 30 22:01 part-00009-cff368dd-78bc-4473-b70d-4037f66bedcd-c000.csv
```

Figure 4-2. *Spark saved data*

The reason is as follows. By default Spark saves DataFrames, datasets, or RDDs in a folder with the name specified and writes the content inside in multiple part files (one part per partition) having the format file specified as extension. As you have seen in the preceding output, Spark also writes a _SUCCESS file and a .crc file for each partition.

If for any reason you require to have the data merged into a single file and get rid of the folder and collateral files, you can only easily achieve the first one of your wishes.

As mentioned before, Spark creates a file for each partition. Thus, one way to get a single file is by consolidating all the shuffled data in a single partition using the `coalesce()` and/or `repartition()` method.

NOTICE

Be careful when using the `coalesce()` and/or `repartition()` method with large data volumes as you can overload the driver memory and get into trouble facing OutOfMemory problems.

Let's see how to get a single file with a simple example valid for Scala and PySpark:

```
SpanishWritersAppendDF.coalesce(1)
.write.csv("/Users/aantolinez/Downloads/personas_coalesce.csv")
```

Now, when you look at the output of the preceding code (Figure 4-3), you can see a single CSV file; however, the folder, _SUCCESS file, and .crc hidden files are still there.

```
personas_coalesce.csv:
total 24
drwxr-xr-x@  6 aantolinez  staff   192 Oct 30 22:48 .
drwx------@ 77 aantolinez  staff  2464 Oct 30 22:48 ..
-rw-r--r--   1 aantolinez  staff     8 Oct 30 22:48 ._SUCCESS.crc
-rw-r--r--   1 aantolinez  staff    12 Oct 30 22:48 .part-00000-28aca7b2-f699-4c2a-a710-7fb7bb5227d9-c000.csv.crc
-rw-r--r--   1 aantolinez  staff     0 Oct 30 22:48 _SUCCESS
-rw-r--r--   1 aantolinez  staff    23 Oct 30 22:48 part-00000-28aca7b2-f699-4c2a-a710-7fb7bb5227d9-c000.csv
```

Figure 4-3. *Spark saving to a single file*

For further refinement, such as removing the folder and _SUCCESS and .crc hidden files, you would have to use the Hadoop file system library to manipulate the final output.

Read and Write Apache Parquet Files

Apache Parquet is a free, open source, columnar, and self-describing file format for fast analytical querying. Apache Parquet plays an important role in modern data lakes due to its capabilities to skip irrelevant data permitting efficient queries on large datasets.

Some advantages of the Parquet columnar storage are the following:

- *Columnar*: Parquet is a column-oriented format. In a Paquet file, data is stored as columns, meaning values of each column are stored close to each other, facilitating accessibility and hence querying performance.

- *Self-description*: A Parquet file combines the data itself with the data schema and structure. This combination facilitates the development of tools to read, store, and write Parquet files.

- *Compression*: In a Parquet file, data compression takes place column by column and includes flexible compression options such as extendable encoding schema per data type. It means that we can use different compression encoding according to the data type (long, string, date, etc.) optimizing compression.

- *Performance*: The Parquet format is designed for querying performance. The internal Parquet format structure (which is out of the scope of this book), composed of row groups, header, and footer, minimizes the volume of data read and hence reduces disk I/O. Comparing Parquet with CSV files, the latter must be read in full and uploaded into memory, while the former permits reading only the relevant columns needed to answer our question. The Parquet format allows retrieval of minimum data, implementing vertical and horizontal partitioning of row groups and column chunks as you can see in Figure 4-4. Column chunks are also organized as data pages including metadata information.

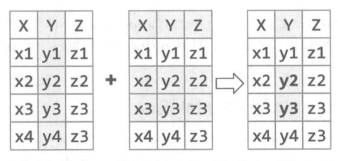

Figure 4-4. *Parquet data partitioning*

Saving and Data Compression of a DataFrame to a Parquet File Format

The parquet() method permits saving a Spark DataFrame to a Parquet file format. By default, this method uses the snappy compression codec:

```
import org.apache.spark.sql.types.{StringType, StructType, IntegerType}

val schemaWriters = new StructType()
```

```
      .add("Name",StringType,true)
      .add("Surname",StringType,true)
      .add("Century",StringType,true)
      .add("YearOfBirth",IntegerType,true)
val SpanishWritersDf = spark.read.option("header", "true")
      .schema(schemaWriters)
.csv("/Users/aantolinez/Downloads/Spanish_Writers_by_Century.csv")
// Saving data with default compression codec: snappy
SpanishWritersDf.write.parquet("/Users/aantolinez/Downloads/Spanish_
Writers_by_Century.parquet")
```

```
Spanish_Writers_by_Century.parquet:
_SUCCESS      part-00000-e4385fd4-fcc0-4a5c-8632-d0080438fa82-c000.gz.parquet
```

The compression codec can be set to none, uncompressed, snappy, gzip, lzo, brotli, lz4, and zstd, overriding the spark.sql.parquet.compression.codec. Data can be appended to a Parquet file using the append option:

```
// Saving data with gzip compression codec compression option
SpanishWritersDf.write.mode("append").option("compression", "gzip").
parquet("/Users/aantolinez/Downloads/Spanish_Writers_by_Century.parquet")
```

As you can see in the following, several compression codecs can be combined:

```
Spanish_Writers_by_Century.parquet:
_SUCCESS      part-00000-e4385fd4-fcc0-4a5c-8632-d0080438fa82-c000.gz.parquet
    part-00000-d070dd4d-86ca-476f-8e67-060365db7ca7-c000.snappy.parquet
```

The same result can be obtained using PySpark code:

```
from pyspark.sql.types import StructField, StringType, StructType,
IntegerType

schemaWriters = StructType([
    StructField("Name",StringType(),True),
    StructField("Surname",StringType(),True),
    StructField("Century",StringType(),True),
    StructField("YearOfBirth", IntegerType(), True)
  ])
```

```
SpanishWritersDf = spark.read.option("header", "true") \
      .schema(schemaWriters) \
.csv("/Users/aantolinez/Downloads/Spanish_Writers_by_Century.csv")
```

Direct Queries on Parquet Files

Spark can run SQL statements directly on Parquet files through temporary views:

```
val parquetDF = spark.read.parquet("/Users/aantolinez/Downloads/Spanish_
Writers_by_Century.parquet")
parquetDF.createOrReplaceTempView("TempTable")
val sqlDf = spark.sql("select * from TempTable where YearOfBirth = 1600")
sqlDf.show()
// Output

+--------+-----------+-------+-----------+
|    Name|    Surname|Century|YearOfBirth|
+--------+-----------+-------+-----------+
|Calderón|de la Barca|   XVII|       1600|
+--------+-----------+-------+-----------+
```

You can get the same result using PySpark code as you see in the following code snippet:

```
parquetDF = spark.read.parquet("/Users/aantolinez/Downloads/Spanish_
Writers_by_Century.parquet")
parquetDF.createOrReplaceTempView("TempTable")
sqlDf = spark.sql("select * from TempTable where YearOfBirth = 1600")
sqlDf.show()
# Output
+--------+-----------+-------+-----------+
|    Name|    Surname|Century|YearOfBirth|
+--------+-----------+-------+-----------+
|Calderón|de la Barca|   XVII|       1600|
+--------+-----------+-------+-----------+
```

Parquet File Partitioning

Spark allows Parquet file partitioning using the partitionBy() method. File partitioning is one of the key Spark features to improve data analytics performance and scalability. File partitioning is a key feature to make reads faster; it allows fast access to the data, loading smaller datasets, and processing data in parallel. Let's see how it works with a small example:

```
import org.apache.spark.sql.types.{StringType, StructType, IntegerType}

val schemaWriters = new StructType()
    .add("Name",StringType,true)
    .add("Surname",StringType,true)
    .add("Century",StringType,true)
    .add("YearOfBirth",IntegerType,true)
    .add("Gender",StringType,true)
val SpanishWritersDf = spark.read.option("header", "true")
    .schema(schemaWriters)
    .csv("/Users/aantolinez/Downloads/Spanish_Writers_by_Gender.csv")
SpanishWritersDf.write.partitionBy("Century","Gender")
.parquet("/Users/aantolinez/Downloads/Spanish_Writers_by_Gender.parquet")
```

Spark creates a folder hierarchy based on "Century" as the first partition key and recursively a group of subfolders for "Gender", the second partition key. You can see the mentioned hierarchy in Figure 4-5.

```
Spanish_Writers_by_Gender.parquet
├── ._SUCCESS.crc
├── Century=XIII
│   └── Gender=M
│       ├── .part-00000-aca04d79-8009-4b12-9e67-45d9f7f975e8.c000.snappy.parquet.crc
│       └── part-00000-aca04d79-8009-4b12-9e67-45d9f7f975e8.c000.snappy.parquet
├── Century=XIV
│   └── Gender=M
│       ├── .part-00000-aca04d79-8009-4b12-9e67-45d9f7f975e8.c000.snappy.parquet.crc
│       └── part-00000-aca04d79-8009-4b12-9e67-45d9f7f975e8.c000.snappy.parquet
├── Century=XIX
│   ├── Gender=F
│   │   ├── .part-00000-aca04d79-8009-4b12-9e67-45d9f7f975e8.c000.snappy.parquet.crc
│   │   └── part-00000-aca04d79-8009-4b12-9e67-45d9f7f975e8.c000.snappy.parquet
│   └── Gender=M
│       ├── .part-00000-aca04d79-8009-4b12-9e67-45d9f7f975e8.c000.snappy.parquet.crc
│       └── part-00000-aca04d79-8009-4b12-9e67-45d9f7f975e8.c000.snappy.parquet
├── Century=XV
│   ├── Gender=F
│   │   ├── .part-00000-aca04d79-8009-4b12-9e67-45d9f7f975e8.c000.snappy.parquet.crc
│   │   └── part-00000-aca04d79-8009-4b12-9e67-45d9f7f975e8.c000.snappy.parquet
│   └── Gender=M
│       ├── .part-00000-aca04d79-8009-4b12-9e67-45d9f7f975e8.c000.snappy.parquet.crc
│       └── part-00000-aca04d79-8009-4b12-9e67-45d9f7f975e8.c000.snappy.parquet
├── Century=XVI
│   ├── Gender=F
│   │   ├── .part-00000-aca04d79-8009-4b12-9e67-45d9f7f975e8.c000.snappy.parquet.crc
│   │   └── part-00000-aca04d79-8009-4b12-9e67-45d9f7f975e8.c000.snappy.parquet
│   └── Gender=M
│       ├── .part-00000-aca04d79-8009-4b12-9e67-45d9f7f975e8.c000.snappy.parquet.crc
│       └── part-00000-aca04d79-8009-4b12-9e67-45d9f7f975e8.c000.snappy.parquet
├── Century=XVII
│   └── Gender=M
│       ├── .part-00000-aca04d79-8009-4b12-9e67-45d9f7f975e8.c000.snappy.parquet.crc
│       └── part-00000-aca04d79-8009-4b12-9e67-45d9f7f975e8.c000.snappy.parquet
├── Century=XX
│   └── Gender=M
│       ├── .part-00000-aca04d79-8009-4b12-9e67-45d9f7f975e8.c000.snappy.parquet.crc
│       └── part-00000-aca04d79-8009-4b12-9e67-45d9f7f975e8.c000.snappy.parquet
└── _SUCCESS
```

Figure 4-5. *Spark Parquet file partitioning by key*

The following PySpark code snippet will return to you the same output:

```
from pyspark.sql.types import StructField, StringType, StructType,
IntegerType

schemaWriters = StructType([
     StructField("Name",StringType(),True),
     StructField("Surname",StringType(),True),
     StructField("Century",StringType(),True),
```

```
        StructField("YearOfBirth", IntegerType(),True),
        StructField("Gender",StringType(),True),
    ])
```

```
SpanishWritersDf = spark.read.option("header", "true") \
        .schema(schemaWriters) \
.csv("/Users/aantolinez/Downloads/Spanish_Writers_by_Gender.csv")
```

```
SpanishWritersDf.write.partitionBy("Century","Gender") \
.parquet("/Users/aantolinez/Downloads/Spanish_Writers_by_Gender.parquet")
```

Reading Parquet File Partitions

One of the ways Spark provides to speed up data processing is by enabling reading only the portions of the data needed.

In the following Scala code snippet, you can see how to select only the data needed from a Parquet file:

```
val partitionDF = spark.read.parquet("/Users/aantolinez/Downloads/Spanish_
Writers_by_Gender.parquet/Century=XX")
partitionDF.show()
+----+---------------+-----------+------+
|Name|        Surname|YearOfBirth|Gender|
+----+---------------+-----------+------+
|José|Ortega y Gasset|       1883|     M|
+----+---------------+-----------+------+
```

The same result is achieved using PySpark code:

```
partitionDF = spark.read.parquet("/Users/aantolinez/Downloads/Spanish_
Writers_by_Gender.parquet/Century=XX")
partitionDF.show()
```

Read and Write JSON Files with Spark

Spark provides two methods for reading JSON files, loading those JSON files as Dataset[Row] and writing data to disk in a JSON format. Spark natively supports JSON file schema deduction, although the JSON file must include separate and valid newline-delimited JSON objects.

You can use `spark.read.json("path/to/file.json")` to load single-line or multiline JSON files or `Dataset[String]` into a Spark DataFrame and use `dataframe.write.json("path/to/file.json")` to save a Spark DataFrame as a JSON file.

Let's see how to load a JSON file into a Spark DataFrame. In the first example, we are going to load a file composed of newline-delimited JSON strings as the one we see in the following:

```
{"id":1,"first_name":"Luis","last_name":"Ortiz","email":"luis.ortiz@mapy.cz",
"country":"Spain","updated":"2015-05-16","registered":false},
{"id":2,"first_name":"Alfonso","last_name":"Antolinez","email":"aantolinez
@optc.es","country":"Spain","updated":"2015-03-11","registered":true},
{"id":3,"first_name":"Juan","last_name":"Dominguez","email":"jdomin@xyz.org",
"country":"Spain","updated":"2015-02-15","registered":true},
{"id":4,"first_name":"Santiago","last_name":"Sanchez","email":"ssanchez
@google.com","country":"Spain","updated":"2014-10-31","registered":false}
```

To load a single and simple JSON file, you can use the `read.json()` as the example shown next:

```
val df = spark.read.json("/Users/aantolinez/Downloads/Spaniards.json")
```

Unlike other data source formats, Spark has the capacity to infer the data schema while reading a JSON file:

```
df.printSchema()
// Output
root
 |-- country: string (nullable = true)
 |-- email: string (nullable = true)
 |-- first_name: string (nullable = true)
 |-- id: long (nullable = true)
 |-- last_name: string (nullable = true)
 |-- registered: boolean (nullable = true)
 |-- updated: string (nullable = true)
```

The same result is obtained when we use PySpark code as shown next:

```
jsonDf = spark.read.json("/Users/aantolinez/Downloads/Spaniards.json")
jsonDf.printSchema()
# Output
root
 |-- country: string (nullable = true)
 |-- email: string (nullable = true)
 |-- first_name: string (nullable = true)
 |-- id: long (nullable = true)
 |-- last_name: string (nullable = true)
 |-- registered: boolean (nullable = true)
 |-- updated: string (nullable = true)
```

However, in many real cases, you are going to find files formatted as arrays of JSON strings. One example of this file format is shown in the following:

```
[{"id":1,"first_name":"Luis","last_name":"Ortiz","email":"luis.ortiz@mapy.
cz","country":"Spain","updated":"2015-05-16","registered":false},
{"id":2,"first_name":"Alfonso","last_name":"Antolinez","email":"aantolinez@
optc.es","country":"Spain","updated":"2015-03-11","registered":true},
{"id":3,"first_name":"Juan","last_name":"Dominguez","email":"jdomin@xyz.org",
"country":"Spain","updated":"2015-02-15","registered":true},
{"id":4,"first_name":"Santiago","last_name":"Sanchez","email":"ssanchez@
google.com","country":"Spain","updated":"2014-10-31","registered":false}]
```

These kinds of files are known as multiline JSON strings. For multiline JSON files, you have to use .option("multiline","true") while reading the data. Let's see how it works with an example in Scala and PySpark:

```
//Loading a multiline JSON strings file into a dataframe. Scala
val multilineJsonDf = spark.read.option("multiline","true")
.json("/Users/aantolinez/Downloads/Spaniards_array.json")
multilineJsonDf.show(4, false)
```

```
// Output
+-------+------------------+----------+---+---------+----------+----------+
|country|email             |first_name|id |last_name|registered|updated   |
+-------+------------------+----------+---+---------+----------+----------+
|Spain  |luis.ortiz@mapy.cz|Luis      |1  |Ortiz    |false     |2015-05-16|
|Spain  |aantolinez@optc.es|Alfonso   |2  |Antolinez|true      |2015-03-11|
|Spain  |jdomin@xyz.org    |Juan      |3  |Dominguez|true      |2015-02-15|
|Spain  |ssanchez@google.com|Santiago |4  |Sanchez  |false     |2014-10-31|
+-------+------------------+----------+---+---------+----------+----------+
```

Reading Multiple JSON Files at Once

The read.json() method can also be used to read multiple files from different paths. To load multiple files at once, you just need to pass their paths as elements of a list. Have a look at how to achieve it in a Scala code snippet:

```scala
//Loading a multiline JSON strings file into a dataframe at once
val multipleJsonsDf = spark.read.option("multiline","true").json(
     "/Users/aantolinez/Downloads/Spaniards_array.json",
     "/Users/aantolinez/Downloads/Spaniards_array2.json")
```

```
// Output
+-------+------------------+----------+---+---------+----------+----------+
|country|email             |first_name|id |last_name|registered|updated   |
+-------+------------------+----------+---+---------+----------+----------+
|Spain  |luis.ortiz@mapy.cz|Luis      |1  |Ortiz    |false     |2015-05-16|
|Spain  |aantolinez@optc.es|Alfonso   |2  |Antolinez|true      |2015-03-11|
|Spain  |jdomin@xyz.org    |Juan      |3  |Dominguez|true      |2015-02-15|
|Spain  |ssanchez@google.com|Santiago |4  |Sanchez  |false     |2014-10-31|
|Spain  |luis.herrera@xyz.es|Luis     |1  |Herrera  |false     |2015-05-15|
|Spain  |mabad@opti.es     |Marcos    |2  |Abad     |true      |2015-03-21|
|Spain  |jabalos@redis.org |Juan      |3  |Abalos   |true      |2015-02-14|
|Spain  |samo@terra.es     |Santiago  |4  |Amo      |false     |2014-10-21|
+-------+------------------+----------+---+---------+----------+----------+
```

The same outcome is achieved using PySpark code:

```
multipleJsonsDf = spark.read.option("multiline","true") \
.json(["/Users/aantolinez/Downloads/Spaniards_array.json", \
      "/Users/aantolinez/Downloads/Spaniards_array2.json"])
multipleJsonsDf.show(10,False)
# Output
```

```
+-------+------------------+----------+---+---------+----------+----------+
|country|email             |first_name|id |last_name|registered|updated   |
+-------+------------------+----------+---+---------+----------+----------+
|Spain  |luis.ortiz@mapy.cz|Luis      |1  |Ortiz    |false     |2015-05-16|
|Spain  |aantolinez@optc.es|Alfonso   |2  |Antolinez|true      |2015-03-11|
|Spain  |jdomin@xyz.org    |Juan      |3  |Dominguez|true      |2015-02-15|
|Spain  |ssanchez@google.com|Santiago |4  |Sanchez  |false     |2014-10-31|
|Spain  |luis.herrera@xyz.es|Luis     |1  |Herrera  |false     |2015-05-15|
|Spain  |mabad@opti.es     |Marcos    |2  |Abad     |true      |2015-03-21|
|Spain  |jabalos@redis.org |Juan      |3  |Abalos   |true      |2015-02-14|
|Spain  |samo@terra.es     |Santiago  |4  |Amo      |false     |2014-10-21|
+-------+------------------+----------+---+---------+----------+----------+
```

Reading JSON Files Based on Patterns at Once

Another command situation you can find in real life is the necessity of reading files
based on name patterns and/or reading all the files in a folder. Spark allows loading
files based on name patterns or the whole files in a directory using the same read.
json() method we have seen in previous examples. Let's see how it works with another
example:

```
val patternJsonsDf = spark.read.option("multiline","true").json(
      "/Users/aantolinez/Downloads/Spaniards_array*.json")
patternJsonsDf.show(20, false)
```

```
// Output
+-------+------------------------+----------+---+---------+----------+----------+
|country|email                   |first_name|id |last_name|registered|updated   |
+-------+------------------------+----------+---+---------+----------+----------+
|Spain  |luis.garcia@xyz.es      |Lucia     |9  |Garcia   |true      |2015-05-15|
|Spain  |maria.rodriguez@opti.es |Maria     |10 |Rodriguez|true      |2015-03-21|
|Spain  |carmen.gonzalez@redis.org|Carmen   |11 |Gonzalez |true      |2015-02-14|
|Spain  |sara.fernandez@terra.es |Sara      |12 |Fernandez|true      |2014-10-21|
|Spain  |luis.ortiz@mapy.cz      |Luis      |1  |Ortiz    |false     |2015-05-16|
|Spain  |aantolinez@optc.es      |Alfonso   |2  |Antolinez|true      |2015-03-11|
|Spain  |jdomin@xyz.org          |Juan      |3  |Dominguez|true      |2015-02-15|
|Spain  |ssanchez@google.com     |Santiago  |4  |Sanchez  |false     |2014-10-31|
|Spain  |luis.herrera@xyz.es     |Luis      |1  |Herrera  |false     |2015-05-15|
|Spain  |mabad@opti.es           |Marcos    |2  |Abad     |true      |2015-03-21|
|Spain  |jabalos@redis.org       |Juan      |3  |Abalos   |true      |2015-02-14|
|Spain  |samo@terra.es           |Santiago  |4  |Amo      |false     |2014-10-21|
+-------+------------------------+----------+---+---------+----------+----------+
```

You can get exactly the same result using PySpark code as follows:

```
patternJsonsDf = spark.read.option("multiline","true").json(
    "/Users/aantolinez/Downloads/Spaniards_array*.json")
patternJsonsDf.show(20, False)
```

In a similar way, you can use patterns to load all the JSON files from a folder. For example, the following code snippets will allow you to read all the JSON files from a directory and only JSON files:

```
// Reading all the JSON files from a directory and only JSON files
in Scala.
val patternJsonsDf = spark.read.option("multiline","true").json(
    "/Users/aantolinez/Downloads/*.json")
```

```
# Reading all the JSON files from a directory and only JSON files in
PySpark.
patternJsonsDf = spark.read.option("multiline","true").json(
    "/Users/aantolinez/Downloads/*.json")
```

Similarly, if you want to read all the files in a directory, you can use the following code:

```scala
//  Reading ALL the files from a directory in Scala.
val patternJsonsDf = spark.read.option("multiline","true").json(
    "/Users/aantolinez/Downloads/")
```

```python
# Reading ALL the files from a directory in PySpark.
patternJsonsDf = spark.read.option("multiline","true").json(
    "/Users/aantolinez/Downloads/")
```

Direct Queries on JSON Files

As with other file formats, like Parquet, Spark also allows us to query JSON files directly. Thus, it is possible to create a SQL query string and pass it to Spark as you will do with a RDBMS.

Suppose you want to directly query the Spaniards.json file shown in previous examples. One way to do it could be by sending the following query to Spark:

```
CREATE TEMPORARY VIEW Spaniards
    USING org.apache.spark.sql.json
    OPTIONS (path '/Users/aantolinez/Downloads/Spaniards.json')
```

As usual, let's see now how to implement it with Scala and PySpark coding:

```scala
// Using Scala code
val sqlContext = new org.apache.spark.sql.SQLContext(sc)
val Spaniards = sqlContext.jsonFile("/Users/aantolinez/Downloads/
Spaniards.json")
Spaniards.registerTempTable("Spaniards")
sqlContext.sql("select * from Spaniards").show(false)
// Output
+-------+-------------------+----------+---+---------+----------+----------+
|country|email              |first_name|id |last_name|registered|updated   |
+-------+-------------------+----------+---+---------+----------+----------+
|Spain  |luis.ortiz@mapy.cz |Luis      |1  |Ortiz    |false     |2015-05-16|
|Spain  |aantolinez@optc.es |Alfonso   |2  |Antolinez|true      |2015-03-11|
|Spain  |jdomin@xyz.org     |Juan      |3  |Dominguez|true      |2015-02-15|
|Spain  |ssanchez@google.com|Santiago  |4  |Sanchez  |false     |2014-10-31|
+-------+-------------------+----------+---+---------+----------+----------+
```

Exactly the same outcome can be achieved using a more compressed code:

```
spark.sqlContext.sql("CREATE TEMPORARY VIEW Spaniards USING json OPTIONS" +
" (path '/Users/aantolinez/Downloads/Spaniards.json')")
spark.sqlContext.sql("select * from Spaniards").show(false)
// Output
+-------+------------------+----------+---+---------+----------+----------+
|country|email             |first_name|id |last_name|registered|updated   |
+-------+------------------+----------+---+---------+----------+----------+
|Spain  |luis.ortiz@mapy.cz|Luis      |1  |Ortiz    |false     |2015-05-16|
|Spain  |aantolinez@optc.es|Alfonso   |2  |Antolinez|true      |2015-03-11|
|Spain  |jdomin@xyz.org    |Juan      |3  |Dominguez|true      |2015-02-15|
|Spain  |ssanchez@google.com|Santiago |4  |Sanchez  |false     |2014-10-31|
+-------+------------------+----------+---+---------+----------+----------+
```

Now we are going to show how to get the same result using PySpark code:

```
# Using PySpark code
spark.sql("CREATE TEMPORARY VIEW Spaniards USING json OPTIONS" + " (path '/
Users/aantolinez/Downloads/Spaniards.json')")
spark.sql("select * from Spaniards").show(10, False)
# Output
+-------+------------------+----------+---+---------+----------+----------+
|country|email             |first_name|id |last_name|registered|updated   |
+-------+------------------+----------+---+---------+----------+----------+
|Spain  |luis.ortiz@mapy.cz|Luis      |1  |Ortiz    |false     |2015-05-16|
|Spain  |aantolinez@optc.es|Alfonso   |2  |Antolinez|true      |2015-03-11|
|Spain  |jdomin@xyz.org    |Juan      |3  |Dominguez|true      |2015-02-15|
|Spain  |ssanchez@google.com|Santiago |4  |Sanchez  |false     |2014-10-31|
+-------+------------------+----------+---+---------+----------+----------+
```

Saving a DataFrame to a JSON File

Apache Spark provides a similar method called `write().json()` to easily save DataFrames to JSON files. The next code snippet shows how to save the multipleJsonsDf dataframe to a permanent storage as a JSON file:

```
multipleJsonsDf.write
 .json("/Users/aantolinez/Downloads/Merged_Spaniards_array.json")
```

Now we check the `Merged_Spaniards_array.json` has been created and split in several partitions as expected:

```
ls Downloads/Merged_Spaniards_array.json
_SUCCESS
part-00000-69975a01-3566-4d2d-898d-cf9e543d81c3-c000.json
part-00001-69975a01-3566-4d2d-898d-cf9e543d81c3-c000.json
```

Saving Modes

As it is with other file formats, the saving modes applicable to JSON files are the same as those shown earlier in Table 4-1. In the next code snippet, you can see how to append data to an already existing JSON file:

```
multipleJsonsDf.write.mode("append").json("/Users/aantolinez/Downloads/
Merged_Spaniards_array.json")
```

```
ls Downloads/Merged_Spaniards_array.json
_SUCCESS
part-00000-188063e9-e5f6-4308-b6e1-7965eaa46c80-c000.json
part-00000-7453b1ad-f3b6-4e68-80eb-254fb539c04d-c000.json
part-00001-188063e9-e5f6-4308-b6e1-7965eaa46c80-c000.json
part-00001-7453b1ad-f3b6-4e68-80eb-254fb539c04d-c000.json
```

The same code can be used for PySpark.

In previous Spark versions, saving modes were identified as SaveMode.ErrorIfExists, SaveMode.Overwrite, SaveMode.Append, and SaveMode.Ignore. However, in the newest Spark releases, the format `mode("errorifexists")`, `mode("append")`, etc. seems to be the way to go.

Load JSON Files Based on Customized Schemas

When we began using JSON files, we said Spark is able to infer the data schema automatically for us when we read a JSON file. However, there are times in which you could be interested in taking advantage of Spark SQL StructType and StructField classes to define your own file schema. This situation could be the case when the data schema you got is too complex for Spark to infer it autonomously:

```
import org.apache.spark.sql.types.{StructType,StructField, StringType,
IntegerType,BooleanType,DateType}
```

```
val schemaSpaniards = StructType(Array(
      StructField("id",StringType,nullable=true),
      StructField("first_name",StringType,nullable=true),
      StructField("last_name",StringType,nullable=true),
      StructField("email", StringType,nullable=true),
      StructField("country", StringType,nullable=true),
      StructField("updated", DateType,nullable=true),
      StructField("registered", BooleanType,nullable=true)
  ))
```

```
val schemaSpaniardsDf = spark.read.schema(schemaSpaniards).json("/Users/
aantolinez/Downloads/Spaniards.json")
```

We can see how the new DataFrame matches the data schema previously defined:

```
schemaSpaniardsDf.printSchema()
// Output
root
 |-- id: string (nullable = true)
 |-- first_name: string (nullable = true)
 |-- last_name: string (nullable = true)
 |-- email: string (nullable = true)
 |-- country: string (nullable = true)
 |-- updated: date (nullable = true)
 |-- registered: boolean (nullable = true)
```

Now we can see the final result after loading the JSON file based on our customized schema:

```
schemaSpaniardsDf.show(false)
// Output
```

```
+---+----------+---------+------------------+-------+----------+----------+
|id |first_name|last_name|email             |country|updated   |registered|
+---+----------+---------+------------------+-------+----------+----------+
|1  |Luis      |Ortiz    |luis.ortiz@mapy.cz |Spain  |2015-05-16|false     |
|2  |Alfonso   |Antolinez|aantolinez@optc.es |Spain  |2015-03-11|true      |
|3  |Juan      |Dominguez|jdomin@xyz.org     |Spain  |2015-02-15|true      |
|4  |Santiago  |Sanchez  |ssanchez@google.com|Spain  |2014-10-31|false     |
+---+----------+---------+------------------+-------+----------+----------+
```

As usual, you get the same result using PySpark code. Let's repeat the previous steps, but this time written in PySpark:

```python
from pyspark.sql.types import StructType,StructField,StringType,
IntegerType,BooleanType,DateType

schemaSpaniards = StructType([ \
StructField("id",IntegerType(),nullable=True), \
StructField("first_name",StringType(),nullable=True), \
StructField("last_name",StringType(),nullable=True), \
StructField("email",StringType(),nullable=True), \
StructField("country",StringType(),nullable=True), \
StructField("updated",DateType(),nullable=True), \
StructField("registered",BooleanType(),nullable=True)])

schemaSpaniardsDf.printSchema()
# Output
root
 |-- id: integer (nullable = true)
 |-- first_name: string (nullable = true)
 |-- last_name: string (nullable = true)
 |-- email: string (nullable = true)
 |-- country: string (nullable = true)
 |-- updated: date (nullable = true)
 |-- registered: boolean (nullable = true)
```

Finally, you can also see the same result, as the one obtained with the Scala script:

```
schemaSpaniardsDf.show(4, False)
# Output
+---+----------+---------+-------------------+-------+----------+----------+
|id |first_name|last_name|email              |country|updated   |registered|
+---+----------+---------+-------------------+-------+----------+----------+
|1  |Luis      |Ortiz    |luis.ortiz@mapy.cz |Spain  |2015-05-16|false     |
|2  |Alfonso   |Antolinez|aantolinez@optc.es |Spain  |2015-03-11|true      |
|3  |Juan      |Dominguez|jdomin@xyz.org     |Spain  |2015-02-15|true      |
|4  |Santiago  |Sanchez  |ssanchez@google.com|Spain  |2014-10-31|false     |
+---+----------+---------+-------------------+-------+----------+----------+
```

Work with Complex Nested JSON Structures Using Spark

In real life, you are barely going to find as simple JSON files as we have shown in previous examples. In particular, if you have to work with NoSQL databases like Apache Cassandra, MongoDB, and others, it is common to find a problem in which nested and complex JSON structures have to be flattened to exchange the data with RDBMS databases as part of an ETL[2] process, to make it more human-readable or facilitate data analytics. Imagine you have a data source with a schema as the one shown just in the following:

```
root
 |-- Book: struct (nullable = true)
 |    |-- Authors: array (nullable = true)
 |    |    |-- element: struct (containsNull = true)
 |    |    |    |-- firstname: string (nullable = true)
 |    |    |    |-- lastname: string (nullable = true)
 |    |-- DOI: string (nullable = true)
 |    |-- Editors: array (nullable = true)
 |    |    |-- element: struct (containsNull = true)
 |    |    |    |-- firstname: string (nullable = true)
 |    |    |    |-- lastname: string (nullable = true)
```

---

[2] ETL (Extract, Transform, Load) is a process to extract, transform, and load data from several sources to a consolidated data repository.

```
|    |-- ISBN: array (nullable = true)
|    |    |-- element: struct (containsNull = true)
|    |    |    |-- Hardcover ISBN: string (nullable = true)
|    |    |    |-- Softcover ISBN: string (nullable = true)
|    |    |    |-- eBook ISBN: string (nullable = true)
|    |-- Id: long (nullable = true)
|    |-- Publisher: string (nullable = true)
|    |-- Title: struct (nullable = true)
|    |    |-- Book Subtitle: string (nullable = true)
|    |    |-- Book Title: string (nullable = true)
|    |-- Topics: array (nullable = true)
|    |    |-- element: string (containsNull = true)
|    |-- eBook Packages: array (nullable = true)
|    |    |-- element: string (containsNull = true)
```

And you would like to transform it into a schema like the one shown in the following:

```
root
 |-- Afirstname: string (nullable = true)
 |-- Alastname: string (nullable = true)
 |-- DOI: string (nullable = true)
 |-- Efirstname: string (nullable = true)
 |-- Elastname: string (nullable = true)
 |-- Hardcover ISBN: string (nullable = true)
 |-- Softcover ISBN: string (nullable = true)
 |-- eBook ISBN: string (nullable = true)
 |-- Id: long (nullable = true)
 |-- Publisher: string (nullable = true)
 |-- Book Subtitle: string (nullable = true)
 |-- Book Title: string (nullable = true)
 |-- Topics: string (nullable = true)
 |-- eBook Packages: string (nullable = true)
```

Hence, flatten the data and get a final Spark DataFrame as the one shown in Figure 4-6.

firstname	Alastname	DOI	firstname	Elastname	Hardcover ISBN	Softcover ISBN	eBook ISBN	Id	Publisher	Book Subtitle	Book Title	Topics	eBook Packages
Adam	Freeman	https://doi.org/1...	null	null	978-1-4842-7354-8	978-1-4842-7355-5	1	Apress Berkeley, CA	The Complete Guid...	Pro Go	Programming Language	Professional and ...
Adam	Freeman	https://doi.org/1...	null	null	978-1-4842-7354-8	978-1-4842-7355-5	1	Apress Berkeley, CA	The Complete Guid...	Pro Go	Programming Language	Professional and ...
Adam	Freeman	https://doi.org/1...	null	null	978-1-4842-7354-8	978-1-4842-7355-5	1	Apress Berkeley, CA	The Complete Guid...	Pro Go	Programming Language	Apress Access Books
Adam	Freeman	https://doi.org/1...	null	null	978-1-4842-7354-8	978-1-4842-7355-5	1	Apress Berkeley, CA	The Complete Guid...	Pro Go	Open Source	Professional and ...
Adam	Freeman	https://doi.org/1...	null	null	978-1-4842-7354-8	978-1-4842-7355-5	1	Apress Berkeley, CA	The Complete Guid...	Pro Go	Open Source	Apress Access Books
Adam	Freeman	https://doi.org/1...	null	null	978-1-4842-7354-8	978-1-4842-7355-5	1	Apress Berkeley, CA	The Complete Guid...	Pro Go	Programming Techn...	Professional and ...
Adam	Freeman	https://doi.org/1...	null	null	978-1-4842-7354-8	978-1-4842-7355-5	1	Apress Berkeley, CA	The Complete Guid...	Pro Go	Programming Techn...	Apress Access Books
Hien	Luu	https://doi.org/1...	null	null	null	978-1-4842-3579-9	2	Apress Berkeley, CA	With Resilient Di...	Beginning Apache	Big Data	Professional and ...
Hien	Luu	https://doi.org/1...	null	null	null	978-1-4842-3579-9	2	Apress Berkeley, CA	With Resilient Di...	Beginning Apache	Big Data	Apress Access Books
Hien	Luu	https://doi.org/1...	null	null	null	978-1-4842-3579-9	2	Apress Berkeley, CA	With Resilient Di...	Beginning Apache	Java	Professional and ...
Hien	Luu	https://doi.org/1...	null	null	null	978-1-4842-3579-9	2	Apress Berkeley, CA	With Resilient Di...	Beginning Apache	Java	Professional and ...

Figure 4-6. Flattened nested JSON file

For this purpose, Spark provides some specific functions such as explode(), to return a new row for each element in a given array or map. This function takes a column name as a parameter containing an array or map of values. When an array is passed, it creates a new column, named "col" by default, containing every element of the array. When this function receives a map, it creates two new columns named "key" and "value" by default and creates a new row for every key and value. However, explode() ignores null or empty values; therefore, if you are interested in these values as well, you should use explode_outer(), which returns null in case the array or map passed is null or empty.

There are other complementary functions you might be interested in exploring, such as posexplode() and posexplode_outer(). The former, apart from creating columns for the elements of an array or map, also creates an additional column named "pos" to hold the position of the array and map elements.

Let's explain with the example shown in Figure 4-6 how some of those functions work. Feel free to uncomment the code lines you find in the following code snippet and run line by line the code, to see the evolution of the schema structure and data:

```
val dfMlBooks = spark.read.option("multiline", "true").json("file:///Users/
aantolinez/Books_array.json")
// dfMlBooks.show(false)
val df2 =dfMlBooks.select("Book.*")
// df2.printSchema()
val df3=df2.select(explode_outer($"Authors"), col("DOI"), $"Editors",
$"ISBN", col("Id"), col("Publisher"), $"Title", col("Topics"), $"eBook
Packages")
// df3.show(false)
// df3.printSchema()
val df4=df3.select(col("col.*"),  col("DOI"), explode_outer($"Editors"),
$"ISBN", col("Id"), col("Publisher"), $"Title", col("Topics"), $"eBook
Packages")
// df4.show(false)
// df4.printSchema()
val df5=df4.select(col("firstname").alias("Afirstname"),col("lastname").
alias("Alastname"),col("DOI"), col("col.*"),explode_outer($"ISBN"),
col("Id"), col("Publisher"), $"Title", col("Topics"), $"eBook Packages")
// df5.show(false)
```

```
// df5.printSchema()
val df6=df5.
select(col("Afirstname"),col("Alastname"),col("DOI"),col("firstname").
alias("Efirstname"),col("lastname").alias("Elastname"),col("col.
Hardcover ISBN").alias("Hardcover ISBN"),col("col.Softcover ISBN").
alias("Softcover ISBN"),col("col.eBook ISBN").alias("eBook ISBN"),
col("Id"), col("Publisher"),col("Title.Book Subtitle").alias("Book
Subtitle"),col("Title.Book Title").alias("Book Title") ,explode_
outer($"Topics").alias("Topics"), $"eBook Packages")
// df6.show(false)
// df6.printSchema()
val df7=df6.
select(col("Afirstname"),col("Alastname"),col("DOI"),col("Efirstname"),
col("Elastname"),col("Hardcover ISBN"),col("Softcover ISBN"),col("eBook
ISBN"),col("Id"),col("Publisher"),col("Book Subtitle"),col("Book
Title"),col("Topics"),explode_outer( $"eBook Packages").alias("eBook
Packages"))
// df7.show(false)
// df7.printSchema()
val df8=df7.select("*")
df8.show(false)
```

The same outcome can be achieved using a similar PySpark code.

Read and Write CSV Files with Spark

Apache Spark SQL provides two specific functions to read and write CSV files. The method spark.read().csv() reads a file or directory of CSV files into a DataFrame. Additionally, dataframe.write().csv() writes a Spark DataFrame to a CSV file.

Spark provides the option() and options() functions to customize the read() and write() behavior. The latter permits specifying several options at once.

We are going to see how to use these four functions mentioned just above can be used to load and write CSV files in the following examples.

The generic use of the read() function could be as follows:

```
val PATH ="Downloads/Spanish_Writers_by_Century_II.csv"
val df0 = spark.read.csv(PATH)
```

```
df0.show(5)
// Output
+--------------------+
|                 _c0|
+--------------------+
|Name;Surname;Cent...|
|Gonzalo;de Berceo...|
| Juan ;Ruiz;XIV;1283|
|Fernando;de Rojas...|
|Garcilaso;de la V...|
+--------------------+
```

Exactly the same output would be achieved if you use PySpark code, as follows:

```
# PySpark version to upload Spanish_Writers_by_Century_II.csv
PATH ="Downloads/Spanish_Writers_by_Century_II.csv"
df0 = spark.read.csv(PATH)
df0.show(5)
```

We can take advantage of the option() function to specify a field's delimiter. The default delimiter is ",":

```
val df1 = spark.read.option("delimiter", ";").csv(PATH)
df1.show(5)
// Output
+---------+----------+-------+-----------+
|      _c0|       _c1|    _c2|        _c3|
+---------+----------+-------+-----------+
|     Name|   Surname|Century|YearOfBirth|
|  Gonzalo| de Berceo|   XIII|       1196|
|    Juan |      Ruiz|    XIV|       1283|
| Fernando|  de Rojas|     XV|       1465|
|Garcilaso|de la Vega|    XVI|       1539|
+---------+----------+-------+-----------+
```

To skip the first line and use it as column names, we can use .option("header", "true"):

```
val df2 = spark.read.option("delimiter", ";").option("header", "true").
csv(path)
df2.show(5)
// Output
+---------+------------+-------+-----------+
|     Name|     Surname|Century|YearOfBirth|
+---------+------------+-------+-----------+
|  Gonzalo|   de Berceo|   XIII|       1196|
|    Juan |        Ruiz|    XIV|       1283|
| Fernando|    de Rojas|     XV|       1465|
|Garcilaso|  de la Vega|    XVI|       1539|
|   Miguel|de Cervantes|    XVI|       1547|
+---------+------------+-------+-----------+
```

Several CSV manipulation options can be specified at once using the options() function:

```
val df3 = spark.read.options(Map("inferSchema"->"true","delimiter"->";",
"header"->"true")).csv(PATH)
df3.show(5)
// Output
+---------+------------+-------+-----------+
|     Name|     Surname|Century|YearOfBirth|
+---------+------------+-------+-----------+
|  Gonzalo|   de Berceo|   XIII|       1196|
|    Juan |        Ruiz|    XIV|       1283|
| Fernando|    de Rojas|     XV|       1465|
|Garcilaso|  de la Vega|    XVI|       1539|
|   Miguel|de Cervantes|    XVI|       1547|
+---------+------------+-------+-----------+
```

Compressed files can also be uploaded using the "compression" option:

```
val GZIP_PATH = "Downloads/Spanish_Writers_by_Century_II.csv.gz"

val df5 = spark.read.option("delimiter", ";").option("header", "true").
option("compression", "gzip").csv(GZIP_PATH)
df5.show(5)
// Output
+---------+------------+-------+----------+
|     Name|     Surname|Century|YearOfBirth|
+---------+------------+-------+----------+

|  Gonzalo|   de Berceo|   XIII|      1196|
|    Juan |        Ruiz|    XIV|      1283|
| Fernando|    de Rojas|     XV|      1465|
|Garcilaso| de la Vega|    XVI|      1539|
|   Miguel|de Cervantes|    XVI|      1547|

+---------+------------+-------+----------+
```

Other important options are nullValue, nanValue, and dateFormat. The first option permits establishing a string representing a null value. The second option permits the specification of a string as representation of a non-number value (NaN by default). The last option sets the string that indicates a date format (by default "yyyy-MM-dd").

To save a Spark DataFrame to a CSV format, we can use the write() function. The write() function takes a folder as a parameter. That directory represents the output path in which the CSV file, plus a _SUCCESS file, will be saved:

```
// To save a DataFrame to a CSV file
OUTPUT_PATH="Downloads/"
df5.write.option("header","true").csv(OUTPUT_PATH)
```

As it happens with other file formats like Parquet, several saving options, Overwrite, Append, Ignore, and the default option ErrorIfExists, are available.

Read and Write Hive Tables

Apache Spark SQL also provides the capability of reading and writing data stored in Apache Hive tables. In this section we are going to show a typical Spark workflow in which we read data from an external source (CSV file) into a Spark DataFrame and save it to a Hive table later on.

The first and second steps you should take are to create a Hive database and table if you do not have them:

```
-- Creating the Hive database we are going to use
CREATE DATABASE IF NOT EXISTS spaniards;
--Creating the Hive table we are going to use
CREATE TABLE IF NOT EXISTS spaniards.writersByCentury (
Name string,
Surname string,
Century string,
YearOfBirth int )
COMMENT 'Spaniards writers by century'
ROW FORMAT DELIMITED
FIELDS TERMINATED BY ',';
```

After creating the basic Hive resources, we can write our code to load the data from a file and save it to the Hive table:

```
import java.io.File
import org.apache.spark.sql.{Row, SaveMode, SparkSession}

val warehouseLocation = "hdfs://localhost:9745/user/hive/warehouse"

val spark = SparkSession
  .builder()
  .appName("Hands-On Spark 3")
  .config("spark.sql.warehouse.dir", warehouseLocation)
  .enableHiveSupport()
  .getOrCreate()

import spark.implicits._
import spark.sql

val path = "file:///tmp/Spanish_Writers_by_Century.csv"

val df = spark.read.option("header", "true").csv(path)
df.show(5,false)
```

```
// Output
+---------+------------+-------+----------+
|Name     |Surname     |Century|YearOfBirth|
+---------+------------+-------+----------+
|Gonzalo  |de Berceo   |XIII   |1196      |
|Juan     |Ruiz        |XIV    |1283      |
|Fernando |de Rojas    |XV     |1465      |
|Garcilaso|de la Vega  |XVI    |1539      |
|Miguel   |de Cervantes|XVI    |1547      |
+---------+------------+-------+----------+
```

```
// Saving now the dataframe to a Hive table
df.write.mode("overwrite").saveAsTable("spaniards.writersByCentury")
```

After saving the data, we can go to our Hive server and check the data is already there:

```
hive> select * from spaniards.writersByCentury;
OK
SLF4J: Failed to load class "org.slf4j.impl.StaticLoggerBinder".
SLF4J: Defaulting to no-operation (NOP) logger implementation
SLF4J: See http://www.slf4j.org/codes.html#StaticLoggerBinder for further
details.
Gonzalo     de Berceo       XIII    1196
Juan        Ruiz            XIV     1283
Fernando    de Rojas        XV      1465
Garcilaso   de la Vega      XVI     1539
Miguel      de Cervantes    XVI     1547
Francisco   de Quevedo      XVI     1580
Luis        de Góngora      XVI     1561
Lope        de Vega         XVI     1562
Tirso       de Molina       XVI     1583
Calderón    de la Barca     XVII    1600
Adolfo      Bécquer         XIX     1836
Benito      Pérez Galdós    XIX     1843
Emilia      Pardo Bazán     XIX     1851
José        Ortega y Gasset XX      1883
Time taken: 9.226 seconds, Fetched: 14 row(s)
hive>
```

Read and Write Data via JDBC from and to Databases

Apache Spark can also write to and read from numerous data sources using a JDBC connector. Something you have to take into consideration while using JDBC connections to access your data is that the information returned from the source is formatted as a Spark DataFrame, which is very convenient.

If you plan to access remote data sources using a JDBC connection, the first thing you need is a JDBC driver compatible with your data source. In the following examples, we are going to walk you through the implementation of JDBC connections to two of the most popular RDBMSs, MySQL and PostgreSQL:

```
val spark = SparkSession.builder
.appName("Hands-On Guide to Apache Spark 3")
.master("local[*]")
.config("spark.driver.memory", "1g")
.config("spark.sql.ansi.enabled ",true)
.config("spark.jars", "./postgresql-42.5.0.jar, ./mysql-connector-
java-8.0.30.jar")
.getOrCreate()

// Connecting to a PostgreSQL remote server
val dfPostgresql = spark.read
.format("jdbc")
.option("url", "jdbc:postgresql://dbserver_url:5432/northwind")
.option("driver", "org.postgresql.Driver")
.option("dbtable","public.categories")
.option("user","YOUR_USER_HERE")
.option("password", "YOUR_PASSWORD_HERE")
.load()

dfPostgresql.show()
// Output
+-----------+--------------+--------------------+-------+
|category_id| category_name|         description|picture|
+-----------+--------------+--------------------+-------+
|          1|     Beverages|Soft drinks, coff...|     []|
|          2|    Condiments|Sweet and savory ...|     []|
```

```
|        3|   Confections|Desserts, candies...|    []|
|        4|Dairy Products|            Cheeses|    []|
|        5|Grains/Cereals|Breads, crackers,...|    []|
|        6|  Meat/Poultry|      Prepared meats|    []|
|        7|       Produce|Dried fruit and b...|    []|
|        8|       Seafood|    Seaweed and fish|    []|
+---------+--------------+--------------------+------+
```

Now we are going to show how to connect to a MySQL database:

```
val jdbcMySQL = spark.read
  .format("jdbc")
  .option("url", "jdbc:mysql://dbserver_url:3306/northwind")
  .option("driver", "com.mysql.jdbc.Driver")
  .option("dbtable", "customers")
  .option("user", "YOUR_USER_HERE")
  .option("password", "YOUR_PASSWORD_HERE")
  .load()

jdbcAwsMySQL.select("id","company","last_name","first_name","job_
title")show(8)
// Output
+---+---------+----------------+----------+--------------------+
| id|  company|       last_name|first_name|           job_title|
+---+---------+----------------+----------+--------------------+
|  1|Company A|          Bedecs|      Anna|               Owner|
|  2|Company B|Gratacos Solsona|   Antonio|               Owner|
|  3|Company C|            Axen|    Thomas|Purchasing Repres...|
|  4|Company D|             Lee| Christina|  Purchasing Manager|
|  5|Company E|        O'Donnell|    Martin|               Owner|
|  6|Company F|     Pérez-Olaeta| Francisco|  Purchasing Manager|
|  7|Company G|             Xie| Ming-Yang|               Owner|
|  8|Company H|        Andersen| Elizabeth|Purchasing Repres...|
+---+---------+----------------+----------+--------------------+
```

You can get the same result using an embedded SQL query instead of retrieving the whole table:

```
val jdbcMySQL = spark.read
  .format("jdbc")
  .option("url", "jdbc:mysql://dbserver_url:3306/northwind")
  .option("driver", "com.mysql.jdbc.Driver")
  .option("query", "select id,company,last_name,first_name,job_title from
  customers")
  .option("user", "YOUR_USER_HERE")
  .option("password", "YOUR_PASSWORD_HERE")
  .load()

jdbcAwsMySQL.show(8)
// Output
+---+---------+----------------+----------+--------------------+
| id|  company|       last_name|first_name|           job_title|
+---+---------+----------------+----------+--------------------+
|  1|Company A|          Bedecs|      Anna|               Owner|
|  2|Company B|Gratacos Solsona|   Antonio|               Owner|
|  3|Company C|            Axen|    Thomas|Purchasing Repres...|
|  4|Company D|             Lee| Christina|  Purchasing Manager|
|  5|Company E|       O'Donnell|    Martin|               Owner|
|  6|Company F|    Pérez-Olaeta| Francisco|  Purchasing Manager|
|  7|Company G|             Xie| Ming-Yang|               Owner|
|  8|Company H|        Andersen| Elizabeth|Purchasing Repres...|
+---+---------+----------------+----------+--------------------+
```

In a similar way, you can save a Spark DataFrame to a database using a JDBC connection. Let's see it with an example of how to add data to a MySQL database table:

```
import spark.implicits._
val data = Seq((6, "Alfonso"))
val dataRdd = spark.sparkContext.parallelize(data)
val dfFromRDD = dataRdd.toDF("id","name")

dfFromRDD.write
  .mode("append")
```

```
.format("jdbc")
.option("url", "jdbc:mysql://dbserver_url:3306/northwind")
.option("driver", "com.mysql.jdbc.Driver")
.option("dbtable", "customers")
.option("user", "YOUR_USER_HERE")
.option("password", "YOUR_PASSWORD_HERE")
.save()
```

There are many other options you can use when using a JDBC connection. The ones we show next are probably the most relevant when you want to optimize the communication with the data source:

```
val jdbcMySQL = spark.read
  .format("jdbc")
  .option("url", "jdbc:mysql://dbserver_url:3306/northwind")
  .option("driver", "com.mysql.jdbc.Driver")
  .option("dbtable", "customers")
  .option("numpartitions", numpartitions)
  .option("lowerbound", min)
  .option("upperbound", max)
  .option("partitioncolumn", primarykey)
  .option("fetchsize", 0)
  .option("batchsize", 1000)
  .option("user", "YOUR_USER_HERE")
  .option("password", "YOUR_PASSWORD_HERE")
  .load()
```

The new options add the following features:

- *numPartitions*: This option is used for both reading and writing and represents the maximum number of partitions used for parallelism processing as well as the maximum number of JDBC connections.

- *partitionColumn*: Represents the column of the table used for partition. It must be of numeric, date, or timestamp format.

- *lowerBound* and *upperBound*: These options are used for reading operations, and they establish the partition stride.

- *fetchsize*: This option is intended to boost JDBC connection performance: it establishes the number of rows to fetch per round trip. The default value is 0.

- *batchsize*: This option only applies to writing operations, and its goal is improving writing performance. It establishes the number of rows to be inserted per round trip.

4.2 Use of Spark DataFrames

The main use of Spark DataFrames is to execute queries. In this section we are going to explore some of the most common operations we can perform by making use of Spark DataFrames such as selection, filtering, aggregations, and data grouping.

In this section we are going to take advantage of the WorldCup[3] dataset to walk you through the use of Spark DataFrames to query data.

Select DataFrame Columns

Probably the most important transformation function you are going to use in Spark is `select()`. This Spark function returns a new Spark DataFrame composed of a selected set of columns. The returned columns can be renamed using the `alias()` function to eliminate ambiguity and/or improve human readability.

The `select()` function takes the name(s) of one or several DataFrame columns and returns a new Spark DataFrame containing only the selected columns. By default Spark does not show the content of the new DataFrame. As you have seen in the book, we can use the `show()` function to instruct Spark to reveal the returned values.

The following code snippet illustrates how to use `show()` to display the data retrieved from a `select()` statement:

```
val dfWC=spark.read.option("header", "true").csv("file:///Users/aantolinez/
Downloads/WorldCups.csv")
dfWC.show(5,false)
```

---

[3] www.kaggle.com/datasets/abecklas/fifa-world-cup?select=WorldCups.csv

```
// Output showing only the first 5 rows
+----+-----------+---------+--------------+-------+----------+-----------+
-------------+-------------+----------+
|Year|Country    |Winner   |Runners-Up    |Third  |Fourth    |GoalsScored|
QualifiedTeams|MatchesPlayed|Attendance|
+----+-----------+---------+--------------+-------+----------+-----------+
-------------+-------------+----------+
|1930|Uruguay    |Uruguay  |Argentina     |USA    |Yugoslavia|70         |
13            |18           |590.549   |
|1934|Italy      |Italy    |Czechoslovakia|Germany|Austria   |70         |
16            |17           |363.000   |
|1938|France     |Italy    |Hungary       |Brazil |Sweden    |84         |
15            |18           |375.700   |
|1950|Brazil     |Uruguay  |Brazil        |Sweden |Spain     |88         |
13            |22           |1.045.246 |
|1954|Switzerland|Germany FR|Hungary      |Austria|Uruguay   |140        |
16            |26           |768.607   |
+----+-----------+---------+--------------+-------+----------+-----------+
-------------+-------------+----------+
only showing top 5 rows
```

The show() function without parameters displays 20 rows and truncates the text length to 20 characters by default. However, show() can take up to three parameters: The first one is an integer corresponding to the number of rows to display. The second parameter can be a Boolean value, indicating whether text string should be truncated, or an integer, denoting the number of characters to display. The third parameter is a Boolean-type value, designating whether values should be shown vertically.

The following output displays the results of the previous example, but using the dfWC.show(5,8) option, showing only the first five rows and just eight characters in length:

```
+----+--------+--------+----------+-------+--------+-----------+
-------------+-------------+----------+
|Year| Country|  Winner|Runners-Up|  Third|  Fourth|GoalsScored|
QualifiedTeams|MatchesPlayed|Attendance|
+----+--------+--------+----------+-------+--------+-----------+
-------------+-------------+----------+
|1930| Uruguay| Uruguay|   Argen...|      USA|Yugos...|        70|
           13|          18|   590.549|
|1934|   Italy|   Italy|   Czech...|Germany| Austria|        70|
           16|          17|   363.000|
|1938|  France|   Italy|   Hungary| Brazil|  Sweden|        84|
           15|          18|   375.700|
|1950|  Brazil| Uruguay|    Brazil| Sweden|   Spain|        88|
           13|          22|   1.045...|
|1954|Switz...|Germa...|   Hungary|Austria| Uruguay|       140|
           16|          26|   768.607|
+----+--------+--------+----------+-------+--------+-----------+
-------------+-------------+----------+
only showing top 5 rows
```

Selecting All or Specific DataFrame Columns

You can use select() to discriminate the columns you would like to query. To retrieve all the columns in the DataFrame, you use show() as explained in the previous examples, or you can use the wildcard "*", as follows:

```
dfWC.select("*").show()
```

On the other hand, you can fetch specific columns from your DataFrame using column names in one of the following query ways:

```
// Fetch specific columns from a DataFrame using column names
dfWC.select("Year","Country", "Winner", "Runners-Up", "Third","Fourth").
show(5, false)
```

```
// Output
+----+-----------+----------+--------------+-------+----------+
|Year|Country    |Winner    |Runners-Up    |Third  |Fourth    |
+----+-----------+----------+--------------+-------+----------+
|1930|Uruguay    |Uruguay   |Argentina     |USA    |Yugoslavia|
|1934|Italy      |Italy     |Czechoslovakia|Germany|Austria   |
|1938|France     |Italy     |Hungary       |Brazil |Sweden    |
|1950|Brazil     |Uruguay   |Brazil        |Sweden |Spain     |
|1954|Switzerland|Germany FR|Hungary       |Austria|Uruguay   |
+----+-----------+----------+--------------+-------+----------+
```

```
// Fetch individual columns from a DataFrame using Dataframe object name
dfWC.select(dfWC("Year"),dfWC("Country"),dfWC("Winner"),dfWC("Runners-Up"),
dfWC("Third"),dfWC("Fourth")).show(5, false)
```

```
//Fetch individual columns from a DataFrame using col function.
import org.apache.spark.sql.functions.col
dfWC.select(col("Year"),col("Country"),col("Winner"),col("Runners-Up"),col(
"Third"),col("Fourth")).show(5, false)
```

You can also select columns from a DataFrame based on column index. You can see a couple of examples in the following:

```
dfWC.select(dfWC.columns(0),dfWC.columns(1),dfWC.columns(2),dfWC.
columns(3)).show(5, false)
// Output
+----+-----------+----------+--------------+
|Year|Country    |Winner    |Runners-Up    |
+----+-----------+----------+--------------+
|1930|Uruguay    |Uruguay   |Argentina     |
|1934|Italy      |Italy     |Czechoslovakia|
|1938|France     |Italy     |Hungary       |
|1950|Brazil     |Uruguay   |Brazil        |
|1954|Switzerland|Germany FR|Hungary       |
+----+-----------+----------+--------------+
```

You can fetch an array of columns using a sequence of indexes:

```
val colIndex = Seq(0, 1, 2, 3, 4, 5)
dfWC.select(colIndex map dfWC.columns map col: _*).show(5, false)
// Output
```

```
+----+-----------+----------+--------------+-------+----------+
|Year|Country    |Winner    |Runners-Up    |Third  |Fourth    |
+----+-----------+----------+--------------+-------+----------+
|1930|Uruguay    |Uruguay   |Argentina     |USA    |Yugoslavia|
|1934|Italy      |Italy     |Czechoslovakia|Germany|Austria   |
|1938|France     |Italy     |Hungary       |Brazil |Sweden    |
|1950|Brazil     |Uruguay   |Brazil        |Sweden |Spain     |
|1954|Switzerland|Germany FR|Hungary       |Austria|Uruguay   |
+----+-----------+----------+--------------+-------+----------+
```

Sequences can also be used in several other ways, for instance, using the sequence plus the string column names, as you see next:

```
val seqColumnas = Seq("Year","Country","Winner","Runners-Up","Third",
"Fourth")
val result = dfWC.select(seqColumnas.head, seqColumnas.tail: _*).
show(5, false)
```

Another way could be using a sequence plus the map function with a set of column names:

```
dfWC.select(seqColumnas.map(i => col(i)): _*).show(5,false)
```

In both examples, you get exactly the same result you got in previous code snippets.

You can also use a list of columns to retrieve the desired data. Have a look at the next example:

```
import org.apache.spark.sql.Column
val miColumnas: List[Column] = List(new Column("Year"), new
Column("Country"), new Column("Winner"))
dfWC.select(miColumnas: _*).show(5,false)
```

```
// Output
+----+-----------+----------+
|Year|Country    |Winner    |
+----+-----------+----------+
|1930|Uruguay    |Uruguay   |
|1934|Italy      |Italy     |
|1938|France     |Italy     |
|1950|Brazil     |Uruguay   |
|1954|Switzerland|Germany FR|
+----+-----------+----------+
```

Select Columns Based on Name Patterns

The startsWith(String prefix) and endsWith(String suffix) column functions are used to confirm whether a string begins with a specified prefix or substring, in the first case, or the same string ends with a defined suffix. Another interesting column function is contains(Other), which returns a Boolean value indicating whether a pattern appears in a column or not. These three functions can be complemented with the like() function to achieve the same results. Let's see how to use them to select the desired columns using name patterns:

```
dfWC.select(dfWC.columns.filter(s=>s.startsWith("Y")).map(c=>col(c)):_*).
show(5,false)
// Output
+----+
|Year|
+----+
|1930|
|1934|
|1938|
|1950|
|1954|
+----+
```

```
dfWC.select(dfWC.columns.filter(s=>s.endsWith("ner")).map(c=>col(c)):_*).
show(5,false)
// Output
+----------+
|Winner    |
+----------+
|Uruguay   |
|Italy     |
|Italy     |
|Uruguay   |
|Germany FR|
+----------+
```

On the other hand, the function contains() can be used to filter rows by columns containing a specific pattern. You can see an example in the following in which we filter the dataset rows with letter "y" in the Winner column:

```
import org.apache.spark.sql.functions.col

dfWC.select("Year","Country","Winner","Runners-Up","Third","Fourth").
filter(col("Winner").contains("S")).show()
// Output
+----+------------+------+----------+-------+-------+
|Year|     Country|Winner| Runners-Up|  Third| Fourth|
+----+------------+------+----------+-------+-------+
|2010|South Africa| Spain|Netherlands|Germany|Uruguay|
+----+------------+------+----------+-------+-------+
```

filter can be complemented with the function like() to achieve the same outcome:

```
dfWC.select("Year","Country","Winner","Runners-Up","Third","Fourth").
filter(col("Winner").like("%S%")).show()
// Outcome
+----+------------+------+----------+-------+-------+
|Year|     Country|Winner| Runners-Up|  Third| Fourth|
+----+------------+------+----------+-------+-------+
|2010|South Africa| Spain|Netherlands|Germany|Uruguay|
+----+------------+------+----------+-------+-------+
```

We can also use SQL ANSI language as a complement to filter dataset rows:

```
dfWC.createOrReplaceTempView("WorldCups")
spark.sql("select Year,Country,Winner,`Runners-Up`,Third,Fourth from
WorldCups where Winner like '%S%'").show()
// Output
+----+------------+------+-----------+-------+-------+
|Year|     Country|Winner| Runners-Up|  Third| Fourth|
+----+------------+------+-----------+-------+-------+
|2010|South Africa| Spain|Netherlands|Germany|Uruguay|
+----+------------+------+-----------+-------+-------+
```

Filtering Results of a Query Based on One or Multiple Conditions

So far we have been applying selection criteria at the column level. Now we are going to see how to refine gathered data at the row level. Spark `filter()` and `where()` functions are used to filter data at the row level based on one or multiple criteria. Both functions return the same outcome; thus, the `where()` function was introduced for SQL background compatibility.

Both filtering functions can be used alone or combined with others to refine the results. In the following code snippet, we are using `filter()` individually; thus, we get the full set of columns. In the second one, we combine it with select to limit the number of columns retrieved to those that are of our interest:

```
dfWC.filter("Year < 1938").show(5,false)
// Output
+----+-------+-------+-------------+-------+----------+-----------+
--------------+--------------+----------+
|Year|Country|Winner |Runners-Up   |Third  |Fourth    |GoalsScored|
QualifiedTeams|MatchesPlayed|Attendance|
+----+-------+-------+-------------+-------+----------+-----------+
--------------+--------------+----------+
|1930|Uruguay|Uruguay|Argentina    |USA    |Yugoslavia|70         |
13            |18            |590.549   |
```

```
|1934|Italy  |Italy  |Czechoslovakia|Germany|Austria    |70         |
16            |17              |363.000   |
+----+-------+-------+--------------+-------+----------+----------+
-------------+------------+----------+
dfWC.select(col("Year"),col("Country"),col("Winner"),col("Runners-Up"),
col("Third"),col("Fourth")).filter("Year < 1938").show(5,false)
// Output
+----+-------+-------+--------------+-------+----------+
|Year|Country|Winner |Runners-Up    |Third  |Fourth    |
+----+-------+-------+--------------+-------+----------+
|1930|Uruguay|Uruguay|Argentina     |USA    |Yugoslavia|
|1934|Italy  |Italy  |Czechoslovakia|Germany|Austria   |
+----+-------+-------+--------------+-------+----------+
```

In the last piece of code, you can appreciate how Spark performs query operations. First, Spark performs the select() transformation, and then it applies the filter criteria to the selected data. Finally, it applies the action show() to the results.

Several filter() functions can be cascaded to provide additional data refinement:

```
dfWC.select(col("Year"),col("Country"),col("Winner"),col("Runners-Up")).
filter("Year < 1938").filter("Country = 'Italy'").show(5,false)
// Output
+----+-------+------+--------------+
|Year|Country|Winner|Runners-Up    |
+----+-------+------+--------------+
|1934|Italy  |Italy |Czechoslovakia|
+----+-------+------+--------------+
```

Using Different Column Name Notations

Column names inside a filter() function can also be mentioned by using 'columnName, col(columnName), $"columnName", Dataframe("columnName") notations. Let's see how it works with an example:

```
// Using the 'columnName notation
dfWC.select("Year","Country","Winner","Runners-Up","Third","Fourth",
"GoalsScored").filter('Winner === "Spain").show(false)
```

```
// Using the $"columnName" notation
dfWC.select("Year","Country","Winner","Runners-Up","Third","Fourth",
"GoalsScored").filter($"Winner" === "Spain").show(false)
// Using the col(columnName) notation
dfWC.select("Year","Country","Winner","Runners-Up","Third","Fourth",
"GoalsScored").filter(col("Winner") === "Winner").show(false)
// Using the Dataframe("columnName") notation
dfWC.select("Year","Country","Winner","Runners-Up","Third","Fourth",
"GoalsScored").filter(dfWC("Winner") === "Spain").show(false)
// Output from all code snippets
+----+------------+------+-----------+-------+-------+-----------+
|Year|Country     |Winner|Runners-Up |Third  |Fourth |GoalsScored|
+----+------------+------+-----------+-------+-------+-----------+
|2010|South Africa|Spain |Netherlands|Germany|Uruguay|145        |
+----+------------+------+-----------+-------+-------+-----------+
```

Alternatively, you can use the whether() function to get the same outcome, as you can see in the following example:

```
// Using the 'columnName notation
dfWC.select("Year","Country","Winner","Runners-Up","Third","Fourth",
"GoalsScored").where('Winner === "Spain").show(false)
// Using the $"columnName" notation
dfWC.select("Year","Country","Winner","Runners-Up","Third","Fourth",
"GoalsScored").where($"Winner" === "Spain").show(false)
// Using the col(columnName) notation
dfWC.select("Year","Country","Winner","Runners-Up","Third","Fourth",
"GoalsScored").where(col("Winner") === "Spain").show(false)
// Using the Dataframe("columnName") notation
dfWC.select("Year","Country","Winner","Runners-Up","Third","Fourth",
"GoalsScored").where(dfWC("Winner") === "Spain").show(false)
// Output from all code snippets
+----+------------+------+-----------+-------+-------+-----------+
|Year|Country     |Winner|Runners-Up |Third  |Fourth |GoalsScored|
+----+------------+------+-----------+-------+-------+-----------+
|2010|South Africa|Spain |Netherlands|Germany|Uruguay|145        |
+----+------------+------+-----------+-------+-------+-----------+
```

Using Logical Operators for Multi-condition Filtering

So far we have shown how to filter dataset rows applying one condition. However, in real life, very often we need more complex selection criteria. The power of the `filter()` function can be enhanced by applying logical operators like AND, OR, and NOT, which will allow you to concatenate multiple conditions. These logical operators can also be represented as "&&", "||", and "!". Once again, we are going to show you how to use them with several examples:

```
// Using AND, "&&" logical operator
dfWC.select("Year","Country","Winner","Runners-Up","Third","Fourth",
"GoalsScored").filter(dfWC("Winner") === "Spain" and dfWC("Runners-Up") ===
"Netherlands").show(false)
// Output
```

Year	Country	Winner	Runners-Up	Third	Fourth	GoalsScored
2010	South Africa	Spain	Netherlands	Germany	Uruguay	145

```
// Using OR, "||" logical operator
dfWC.select("Year","Country","Winner","Runners-Up","Third","Fourth",
"GoalsScored").filter(dfWC("Winner") === "Spain" or dfWC("Runners-Up") ===
"Netherlands").show(false)
// Output
```

Year	Country	Winner	Runners-Up	Third	Fourth	GoalsScored
1974	Germany	Germany FR	Netherlands	Poland	Brazil	97
1978	Argentina	Argentina	Netherlands	Brazil	Italy	102
2010	South Africa	Spain	Netherlands	Germany	Uruguay	145

It seems the Dutch are real experts in being the second one!

One possible use of NOT or "!" could be something like the next one:

```
// Using NOT, "!" logical operator
dfWC.select("Year","Country","Winner","Runners-Up","Third","Fourth").filter
(not (dfWC("Winner") === "Italy" or dfWC("Runners-Up") === "Netherlands")).
show(false)
// Output
```

Year	Country	Winner	Runners-Up	Third	Fourth
1930	Uruguay	Uruguay	Argentina	USA	Yugoslavia
1950	Brazil	Uruguay	Brazil	Sweden	Spain
1954	Switzerland	Germany FR	Hungary	Austria	Uruguay
1958	Sweden	Brazil	Sweden	France	Germany FR
1962	Chile	Brazil	Czechoslovakia	Chile	Yugoslavia
1966	England	England	Germany FR	Portugal	Soviet Union
1970	Mexico	Brazil	Italy	Germany FR	Uruguay
1986	Mexico	Argentina	Germany FR	France	Belgium
1990	Italy	Germany FR	Argentina	Italy	England
1994	USA	Brazil	Italy	Sweden	Bulgaria
1998	France	France	Brazil	Croatia	Netherlands
2002	Korea/Japan	Brazil	Germany	Turkey	Korea Republic
2014	Brazil	Germany	Argentina	Netherlands	Brazil

Manipulating Spark DataFrame Columns

When you will be working in real environments, you are very likely to manipulate the original DataFrame, adding columns, deleting columns, and so on. Apache Spark provides the withColumn() transformation function to manipulate Spark DataFrame columns such as adding a new column, updating the value of a column, changing a column data type, creating an inherited column from existing ones, etc.

The transformation performed with the withColumn() function can be applied to all DataFrame rows or a set of them. As we have just mentioned, withColumn() is a transformation function, and as we have already described, DataFrame transformations return new DataFrames and are lazily evaluated.

Let's see how to take advantage of withColumn() with some examples, as usual. In our first example, we are going to use the previous FIFA World Cup dataset, to add a new calculated column reflecting an important missing metric, the fans attendance per match played.

If you try to immediately apply withColumn() to add a new calculated column, you are going to run into trouble. Why? Have a look at the data schema you got:

```
dfWC.printSchema()
// Output
root
 |-- Year: string (nullable = true)
 |-- Country: string (nullable = true)
 |-- Winner: string (nullable = true)
 |-- Runners-Up: string (nullable = true)
 |-- Third: string (nullable = true)
 |-- Fourth: string (nullable = true)
 |-- GoalsScored: string (nullable = true)
 |-- QualifiedTeams: string (nullable = true)
 |-- MatchesPlayed: string (nullable = true)
 |-- Attendance: string (nullable = true)
```

Do you see the problem? Yes, Spark identified all the columns as string. Therefore, if you attempt to perform the operation at this stage, you will get the following errors:

```
// Adding a new column to DataFrame. Fans attendance per match played
import org.apache.spark.sql.functions.col
val dfWCExt=dfWC.withColumn("AttendancePerMatch", round(col("Attendance")/
col("MatchesPlayed"), 3))
dfWCExt.select("Attendance","MatchesPlayed","AttendancePerMatch").show(5)
// Output
+----------+-------------+------------------+
|Attendance|MatchesPlayed|AttendancePerMatch|
+----------+-------------+------------------+
|   590.549|           18|            32.808|
|   363.000|           17|            21.353|
|   375.700|           18|            20.872|
```

```
| 1.045.246|           22|             null|
|   768.607|           26|           29.562|
+----------+------------+-----------------+
```

Thus, the first step should be converting the columns of interest to a numeric data type. However, if you try to directly convert some figures such as 1.045.246 to numeric, you will also have problems, as they are saved in Metric System format. Therefore, it could be preferable to remove " . " to avoid problems:

```scala
import org.apache.spark.sql.functions.regexp_replace
import org.apache.spark.sql.functions.col
import org.apache.spark.sql.types.IntegerType

val dfWC2=dfWC.withColumn("Attendance", regexp_replace($"Attendance",
"\\.", ""))
val dfWC3=dfWC2.withColumn("GoalsScored", col("GoalsScored").
cast(IntegerType))
.withColumn("QualifiedTeams", col("QualifiedTeams").cast(IntegerType))
.withColumn("MatchesPlayed", col("MatchesPlayed").cast(IntegerType))
.withColumn("Attendance", col("Attendance").cast(IntegerType))
dfWC3.printSchema()
// Output
root
 |-- Year: string (nullable = true)
 |-- Country: string (nullable = true)
 |-- Winner: string (nullable = true)
 |-- Runners-Up: string (nullable = true)
 |-- Third: string (nullable = true)
 |-- Fourth: string (nullable = true)
 |-- GoalsScored: integer (nullable = true)
 |-- QualifiedTeams: integer (nullable = true)
 |-- MatchesPlayed: integer (nullable = true)
 |-- Attendance: integer (nullable = true)
```

Now, we can accomplish our goal of adding a new calculated column:

```
val dfWCExt=dfWC3.withColumn("AttendancePerMatch", round(col("Attendance").
cast(IntegerType)/col("MatchesPlayed").cast(IntegerType), 3))
dfWCExt.select("Attendance","MatchesPlayed","AttendancePerMatch").show(5)
// Output
+----------+-------------+------------------+
|Attendance|MatchesPlayed|AttendancePerMatch|
+----------+-------------+------------------+
|    590549|           18|         32808.278|
|    363000|           17|         21352.941|
|    375700|           18|         20872.222|
|   1045246|           22|         47511.182|
|    768607|           26|         29561.808|
+----------+-------------+------------------+
```

In the preceding example, you have already used a good bunch of the `withColumn()` use cases.

Renaming DataFrame Columns

Apache Spark provides the `withColumnRenamed()` transformation function to change DataFrame column names; it can be used to rename a single column or multiple ones at the same time. This function can also be used to rename nested StructType schemas. The `withColumnRenamed()` function takes two parameters: the current name of the DataFrame column we would like to change and the new name we would like to give to that column. As `withColumnRenamed()` is a transformation, it returns a DataFrame.

Let's see how it works with a practical example. The next code snippet renames one of the dfWCExt columns from AttendancePerMatch to AxMatch:

```
val dfRenamed = dfWCExt.withColumnRenamed("AttendancePerMatch","AxMatch")
dfRenamed.select("Attendance", "MatchesPlayed","AxMatch").show(5)
// Output
+----------+-------------+---------+
|Attendance|MatchesPlayed|  AxMatch|
+----------+-------------+---------+
|    590549|           18|32808.278|
```

```
|      363000|            17|21352.941|
|      375700|            18|20872.222|
|     1045246|            22|47511.182|
|      768607|            26|29561.808|
+----------+-------------+---------+
```

Several columns can be renamed at once joining several calls to withColumnRenamed() as shown next:

```
val dfRenamed2 = dfRenamed
.withColumnRenamed("Attendance","Att")
.withColumnRenamed("AxMatch","AttendancexMatch")
dfRenamed2.select("Att", "MatchesPlayed","AttendancexMatch").show(5)
// Output
+-------+-------------+----------------+
|    Att|MatchesPlayed|AttendancexMatch|
+-------+-------------+----------------+
| 590549|           18|       32808.278|
| 363000|           17|       21352.941|
| 375700|           18|       20872.222|
|1045246|           22|       47511.182|
| 768607|           26|       29561.808|
+-------+-------------+----------------+
```

Dropping DataFrame Columns

Apache Spark provides the drop() function to drop DataFrame columns. The drop() function can be used to remove one column or several at once.

Spark drop() can be used by employing three different syntaxes:

- *Deleting a single column*: dataframe.drop('column name')

- *Deleting multiple columns*: dataframe.drop(*('column 1', 'column 2', 'column n'))

- *Deleting all columns of a dataframe*: dataframe.drop(*list_of_ column names)

Now we are going to see how to use them with several examples continuing with the dataframes we created in the previous sections.

Let's start by droping one of the columns (*AttendancexMatch*) added before:

```
val dfFropOne = dfRenamed2.drop("AttendancexMatch")
dfFropOne.show(5)
// Output
+----+-----------+----------+-------------+-------+----------+-----------+
-------------+------------+-------+
|Year|    Country|    Winner|    Runners-Up|  Third|    Fourth|GoalsScored|
QualifiedTeams|MatchesPlayed|    Att|
+----+-----------+----------+-------------+-------+----------+-----------+
-------------+------------+-------+
|1930|    Uruguay|   Uruguay|     Argentina|    USA|Yugoslavia|         70|
            13|           18| 590549|
|1934|      Italy|     Italy|Czechoslovakia|Germany|   Austria|         70|
            16|           17| 363000|
|1938|     France|     Italy|       Hungary| Brazil|    Sweden|         84|
            15|           18| 375700|
|1950|     Brazil|   Uruguay|        Brazil| Sweden|     Spain|         88|
            13|           22|1045246|
|1954|Switzerland|Germany FR|       Hungary|Austria|   Uruguay|        140|
            16|           26| 768607|
+----+-----------+----------+-------------+-------+----------+-----------+
-------------+------------+-------+
```

After that we are going to drop columns "MatchesPlayed" and "AttendancexMatch":

```
val dfFropTwo = dfRenamed2.drop("MatchesPlayed","AttendancexMatch")
dfFropTwo.show(5)
// Output
+----+-----------+----------+-------------+-------+----------+-----------+
-------------+-------+
|Year|    Country|    Winner|    Runners-Up|  Third|    Fourth|GoalsScored|
QualifiedTeams|    Att|
+----+-----------+----------+-------------+-------+----------+-----------+
-------------+-------+
```

```
|1930|    Uruguay|    Uruguay|    Argentina|      USA|Yugoslavia|        70|
           13|  590549|
|1934|      Italy|      Italy|Czechoslovakia|Germany|    Austria|        70|
           16|  363000|
|1938|     France|      Italy|       Hungary|  Brazil|    Sweden|        84|
           15|  375700|
|1950|     Brazil|    Uruguay|        Brazil|  Sweden|     Spain|        88|
           13|1045246|
|1954|Switzerland|Germany FR|       Hungary| Austria|   Uruguay|       140|
           16|  768607|
+----+-----------+----------+--------------+-------+----------+-----------+
-------------+-------+
```

The same result can be obtained using PySpark code:

```
dfFropTwo = dfRenamed2.drop(*("MatchesPlayed","AttendancexMatch"))
dfFropTwo.show(5)
# Output
+----+-----------+----------+--------------+-------+----------+-----------+
-------------+-------+
|Year|    Country|    Winner|    Runners-Up|  Third|    Fourth|GoalsScored|
QualifiedTeams|    Att|
+----+-----------+----------+--------------+-------+----------+-----------+
-------------+-------+
|1930|    Uruguay|    Uruguay|    Argentina|      USA|Yugoslavia|        70|
           13|  590549|
|1934|      Italy|      Italy|Czechoslovakia|Germany|    Austria|        70|
           16|  363000|
|1938|     France|      Italy|       Hungary|  Brazil|    Sweden|        84|
           15|  375700|
|1950|     Brazil|    Uruguay|        Brazil|  Sweden|     Spain|        88|
           13|1045246|
|1954|Switzerland|Germany FR|       Hungary| Austria|   Uruguay|       140|
           16|  768607|
+----+-----------+----------+--------------+-------+----------+-----------+
-------------+-------+
```

You can delete all the DataFrame columns at once using the following Scala code snippet:

```
val allColumnsList = dfRenamed2.columns
val dfFropAll = dfRenamed2.drop(allColumnsList:_*)
dfFropAll.show(5)
// Output
++
||
++
||
++
```

And the same result is obtained using PySpark:

```
allColumnsList = dfRenamed2.columns
dfFropAll = dfRenamed2.drop(*allColumnsList)
dfFropAll.show(2)
# Output
++
||
++
||
||
++
```

Creating a New Dataframe Column Dependent on Another Column

Spark SQL "case when" and "when otherwise" permit replicating the SQL CASE statement in Spark.

Consider the following dataset:

```
import org.apache.spark.sql.functions.{when, _}
import spark.sqlContext.implicits._

val p = List(("Juan ","Bravo",67,"M",65000),
             ("Miguel ","Rosales",40,"M",87000),
```

```
                ("Roberto ","Morales",7,"M",0),
                ("Maria ","Gomez",12,"F",0),
                ("Vanesa","Lopez",25,"F",72000))
val c = Seq("name","surname","age","gender","salary")
val df = spark.createDataFrame(p).toDF(c:_*)
df.show(5)
// Output
+--------+-------+---+------+------+
|    name|surname|age|gender|salary|
+--------+-------+---+------+------+
|   Juan | Bravo| 67|     M| 65000|
| Miguel |Rosales| 40|     M| 87000|
|Roberto |Morales|  7|     M|     0|
|   Maria | Gomez| 12|     F|     0|
|  Vanesa|  Lopez| 25|     F| 72000|
+--------+-------+---+------+------+
```

Now we are going to see how to implement when() and otherwise() in Scala:

```
val df2 = df
     .withColumn("stage", when(col("age") < 10,"Child")
     .when(col("age") >= 10 && col("age") < 18,"Teenager")
     .otherwise("Adult"))
df2.show(5)
// Output
+--------+-------+---+------+------+--------+
|    name|surname|age|gender|salary|   stage|
+--------+-------+---+------+------+--------+
|   Juan | Bravo| 67|     M| 65000|   Adult|
| Miguel |Rosales| 40|     M| 87000|   Adult|
|Roberto |Morales|  7|     M|     0|   Child|
|   Maria | Gomez| 12|     F|     0|Teenager|
|  Vanesa|  Lopez| 25|     F| 72000|   Adult|
+--------+-------+---+------+------+--------+
```

The when() clause can also be used as part of a SQL select statement:

```
val df3 = df.select(col("*"),
    expr("case when age < '10' then 'Child' " +
        "when age >= '10' and age <= '18' then 'Teenager' "
      + "else 'Adult' end").alias("stage"))
df3.show()
// Output
```

```
+--------+-------+---+------+------+--------+
|    name|surname|age|gender|salary|   stage|
+--------+-------+---+------+------+--------+
|   Juan |  Bravo| 67|     M| 65000|   Adult|
| Miguel |Rosales| 40|     M| 87000|   Adult|
|Roberto |Morales|  7|     M|     0|   Child|
|   Maria|  Gomez| 12|     F|     0|Teenager|
|  Vanesa|  Lopez| 25|     F| 72000|   Adult|
+--------+-------+---+------+------+--------+
```

Here's using the when() clause with null values:

```
val personas = sc.parallelize(Seq(
    ("Juan ","Bravo",67,"M",new Integer(65000) ),
    ("Miguel ","Rosales",40,"M",new Integer(87000) ),
    ("Roberto ","Morales",7,"M",null.asInstanceOf[Integer] ),
    ("Maria ","Gomez",12,"F",null.asInstanceOf[Integer] ),
    ("Vanesa","Lopez",25,"F",new Integer(32000) ))
)
val dfp = personas.toDF("name","surname","age","gender","salary")

// ppower --> purchasing power
dfp.withColumn("ppower", when(col("salary") < 40000,"Low")
   .when(col("salary") >= 40000 && col("Salary") < 70000,"Medium")
   .when(col("salary").isNull ,"")
   .otherwise("High")).show()
// Output
```

```
+--------+-------+---+------+------+------+
|    name|surname|age|gender|salary|ppower|
+--------+-------+---+------+------+------+
|    Juan|  Bravo| 67|     M| 65000|Medium|
|  Miguel|Rosales| 40|     M| 87000|  High|
| Roberto|Morales|  7|     M|  null|      |
|   Maria|  Gomez| 12|     F|  null|      |
|  Vanesa|  Lopez| 25|     F| 32000|   Low|
+--------+-------+---+------+------+------+
```

User-Defined Functions (UDFs)

User-defined functions are a Spark feature designed to help users extend the system's built-in functionalities by writing custom functions.

In this section we are going to use a UDF to categorize the previous personas dataframe between adults and not adults. To create a UDF, the first step is to write a function. In Scala, this kind of functions do not include the return statement, can receive multiple parameters, and do not accept null values. Let's code a simple function that takes an integer value and returns a string, classifying each individual of our dataset between "Adult" or "No adult" depending on their age:

```scala
def  isAdult= (age: Integer) => {
  if(age >= 18){
      "Adult"
  }
    else{
    "No adult"
    }
}
```

The isAdult() function is ready to be used; however, it needs to be registered before it is called. After registering a UDF on the driver node, Spark transfers it over to all executor processes, making it available to all worker machines. In the following we proceed to register the isAdult() function:

```scala
val isAdultUDF = udf(isAdult)
```

175

Now that our UDF is registered, we can use it with our personas dataframe as a normal SQL function:

```
val finalDF=df.withColumn("is_adult",isAdultUDF(col("age")))
finalDF.show()
// Output
+--------+-------+---+------+------+--------+
|    name|surname|age|gender|salary|is_adult|
+--------+-------+---+------+------+--------+
|    Juan|  Bravo| 67|     M| 65000|   Adult|
|  Miguel|Rosales| 40|     M| 87000|   Adult|
| Roberto|Morales|  7|     M|     0|No adult|
|   Maria|  Gomez| 12|     F|     0|No adult|
|  Vanesa|  Lopez| 25|     F| 72000|   Adult|
+--------+-------+---+------+------+--------+
```

Merging DataFrames with Union and UnionByName

Very often in real life, you will have a set of data files that you would like to merge into a single DataFrame. A typical example is when you receive several files containing time-series data. In this case, all the files you receive will have the same data structure. In this scenario, the Spark union() method can be used to merge several DataFrames with the same schema. The union() method has an important limitation. It works by combining the DataFrames by position. This means both DataFrames' columns have to be in the same order; if they are not, the resultant DataFrame will not be correct. If your dataframes do not have the same structure, union() returns an error.

Let's see how to use the union() function with a practical example in which we are going to use two files containing production of crude oil in thousands of barrels:

```
val dfCOP3=spark.read.option("header", "true").csv("file:///Users/
aantolinez/Downloads/Crude_Oil_Production_3.csv")
dfCOP3.show(5)
// Output
```

```
+----+------+------+------+------+------+------+------+------+------+
------+------+------+
|Year|   Jan|   Feb|   Mar|   Apr|   May|   Jun|   Jul|   Aug|   Sep|
   Oct|   Nov|   Dec|
+----+------+------+------+------+------+------+------+------+------+
------+------+------+
|2010|167529|155496|170976|161769|167427|161385|164234|168867|168473|
174547|167272|173831|
|2011|170393|151354|174158|166858|174363|167673|168635|175618|168411|
182977|181157|189487|
|2012|191472|181783|196138|189601|197456|188262|199368|196867|197942|
216057|212472|220282|
|2013|219601|200383|223683|221242|227139|218355|233210|233599|235177|
240600|237597|245937|
|2014|250430|228396|257225|255822|268025|262291|274273|276909|272623|
287256|279821|296518|
+----+------+------+------+------+------+------+------+------+------+
------+------+------+

val dfCOP4=spark.read.option("header", "true").csv("file:///Users/
aantolinez/Downloads/Crude_Oil_Production_4.csv")
dfCOP4.show(5)
// Output
+----+------+------+------+------+------+------+------+------+------+
------+------+------+
|Year|   Jan|   Feb|   Mar|   Apr|   May|   Jun|   Jul|   Aug|   Sep|
   Oct|   Nov|   Dec|
+----+------+------+------+------+------+------+------+------+------+
------+------+------+
|2015|290891|266154|297091|289755|293711|280734|292807|291702|284406|
291419|279982|287533|
|2016|285262|262902|282132|266219|273875|260284|268526|269386|256317|
272918|267097|273288|
|2017|275117|255081|284146|273041|284727|273321|286657|286759|285499|
299726|302564|309486|
```

```
|2018|310032|287870|324467|314996|323491|319216|337814|353154|343298|
356767|356583|370284|
|2019|367924|326845|369292|364458|376763|366546|368965|387073|377710|
397094|390010|402314|
+----+------+------+------+------+------+------+------+------+------+
------+------+------+
```

Merging DataFrames with Duplicate Values

Unlike other SQL functions, Spark union() does not drop duplicate values after combining the DataFrames. If you do not want to have duplicate values in your final DataFrame, you can remove them after merging the DataFrames using the distinct() function. The distinct() function filters duplicate values. Next, there is an example of how to remove duplicate records.

In this example, we would like to merge dfCOP5 and dfCOP4 DataFrames. As you can see in the following, both dfCOP5 and dfCOP4 DataFrames include records of the year 2015, therefore resulting in duplicate 2015 rows in the combined DataFrame:

```
// dfCOP5 DataFrame including Year 2015 records
val dfCOP5=spark.read.option("header", "true").csv("file:///Users/
aantolinez/Downloads/Crude_Oil_Production_5.csv")
dfCOP5.show(10)
// Output
+----+------+------+------+------+------+------+------+------+------+
------+------+------+
|Year|   Jan|   Feb|   Mar|   Apr|   May|   Jun|   Jul|   Aug|   Sep|
   Oct|   Nov|   Dec|
+----+------+------+------+------+------+------+------+------+------+
------+------+------+
|2010|167529|155496|170976|161769|167427|161385|164234|168867|168473|
174547|167272|173831|
|2011|170393|151354|174158|166858|174363|167673|168635|175618|168411|
182977|181157|189487|
|2012|191472|181783|196138|189601|197456|188262|199368|196867|197942|
216057|212472|220282|
|2013|219601|200383|223683|221242|227139|218355|233210|233599|235177|
240600|237597|245937|
```

178

```
|2014|250430|228396|257225|255822|268025|262291|274273|276909|272623|
287256|279821|296518|
|2015|290891|266154|297091|289755|293711|280734|292807|291702|284406|
291419|279982|287533|
+----+------+------+------+------+------+------+------+------+------+
------+------+------+
```

To produce a clear DataFrame, we can use the following code snippet. You can see in the following only one 2015 row is preserved:

```
val cleanDf = dfCOP4.union(dfCOP5).distinct()
cleanDf.show(false)
// Output
+----+------+------+------+------+------+------+------+------+------+
------+------+------+
|Year|Jan   |Feb   |Mar   |Apr   |May   |Jun   |Jul   |Aug   |Sep   |
Oct   |Nov   |Dec   |
+----+------+------+------+------+------+------+------+------+------+
------+------+------+
|2019|367924|326845|369292|364458|376763|366546|368965|387073|377710|
397094|390010|402314|
|2016|285262|262902|282132|266219|273875|260284|268526|269386|256317|
272918|267097|273288|
|2015|290891|266154|297091|289755|293711|280734|292807|291702|284406|
291419|279982|287533|
|2018|310032|287870|324467|314996|323491|319216|337814|353154|343298|
356767|356583|370284|
|2017|275117|255081|284146|273041|284727|273321|286657|286759|285499|
299726|302564|309486|
|2012|191472|181783|196138|189601|197456|188262|199368|196867|197942|
216057|212472|220282|
|2011|170393|151354|174158|166858|174363|167673|168635|175618|168411|
182977|181157|189487|
|2014|250430|228396|257225|255822|268025|262291|274273|276909|272623|
287256|279821|296518|
```

```
|2010|167529|155496|170976|161769|167427|161385|164234|168867|168473|
174547|167272|173831|
|2013|219601|200383|223683|221242|227139|218355|233210|233599|235177|
240600|237597|245937|
+----+------+------+------+------+------+------+------+------+------+
------+------+------+
```

Another very useful Spark method when we want to merge DataFrames is unionByName(). unionByName() permits the combination of several DataFrames by column names instead of by their position; therefore, it is appropriate when DataFrames have the same column names but in different positions.

The unionByName() function since Spark version 3 incorporates allowMissingColumns. When allowMissingColumns is set to true, it allows merging DataFrames when some columns are missing from one DataFrame.

In the following example, we merge DataFrame dfCOP1

```
val dfCOP1=spark.read.option("header", "true").csv("file:///Users/
aantolinez/Downloads/Crude_Oil_Production_1.csv")
dfCOP1.show(5)
// Output
+----+------+------+------+------+------+------+
|Year|   Jan|   Feb|   Mar|   Apr|   May|   Jun|
+----+------+------+------+------+------+------+
|2015|290891|266154|297091|289755|293711|280734|
|2016|285262|262902|282132|266219|273875|260284|
|2017|275117|255081|284146|273041|284727|273321|
|2018|310032|287870|324467|314996|323491|319216|
|2019|367924|326845|369292|364458|376763|366546|
+----+------+------+------+------+------+------+
```

with DataFrame dfCOP2:

```
val dfCOP2=spark.read.option("header", "true").csv("file:///Users/
aantolinez/Downloads/Crude_Oil_Production_2.csv")
dfCOP2.show(5)
// Output
```

```
+----+------+------+------+------+------+------+
|Year|  Jan|   Feb|   Mar|   Apr|   May|   Jun|
+----+------+------+------+------+------+------+
|2015|290891|266154|297091|289755|293711|280734|
|2016|285262|262902|282132|266219|273875|260284|
|2017|275117|255081|284146|273041|284727|273321|
|2018|310032|287870|324467|314996|323491|319216|
|2019|367924|326845|369292|364458|376763|366546|
+----+------+------+------+------+------+------+
```

The dfCOP1 and dfCOP2 DataFrames only have the Year column in common:

```
// Using allowMissingColumns=true
val missingColumnsDf=dfCOP1.unionByName(dfCOP2, allowMissingColumns=true)
missingColumnsDf.show()
// Output
+----+------+------+------+------+------+------+------+------+------+
------+------+------+
|Year|  Jan|   Feb|   Mar|   Apr|   May|   Jun|   Jul|   Aug|   Sep|
   Oct|   Nov|   Dec|
+----+------+------+------+------+------+------+------+------+------+
------+------+------+
|2015|290891|266154|297091|289755|293711|280734|  null|  null|  null|
  null|  null|  null|
|2016|285262|262902|282132|266219|273875|260284|  null|  null|  null|
  null|  null|  null|
|2017|275117|255081|284146|273041|284727|273321|  null|  null|  null|
  null|  null|  null|
|2018|310032|287870|324467|314996|323491|319216|  null|  null|  null|
  null|  null|  null|
|2019|367924|326845|369292|364458|376763|366546|  null|  null|  null|
  null|  null|  null|
|2020|398420|372419|396693|357412|301105|313275|  null|  null|  null|
  null|  null|  null|
|2015|  null|  null|  null|  null|  null|  null|292807|291702|284406|
291419|279982|287533|
```

```
|2016|  null|  null|  null|  null|  null|  null|268526|269386|256317|
272918|267097|273288|
|2017|  null|  null|  null|  null|  null|  null|286657|286759|285499|
299726|302564|309486|
|2018|  null|  null|  null|  null|  null|  null|337814|353154|343298|
356767|356583|370284|
|2019|  null|  null|  null|  null|  null|  null|368965|387073|377710|
397094|390010|402314|
|2020|  null|  null|  null|  null|  null|  null|341184|327875|327623|
324180|335867|346223|
+----+------+------+------+------+------+------+------+------+------+
------+------+------+
```

Spark offers another option to merge DataFrames called unionAll(); however, it is deprecated since Spark 2 in favor of union().

Wrapping up, it is important to underline that union() and unionByName() merge DataFrames vertically on top of each other.

In the next section, we are going to see another way of joining DataFrames.

Joining DataFrames with Join

In the last section, you saw how to glue DataFrames vertically, stacking up one over the other. In this section you are going to see another way of combining DataFrames, but this time horizontally, one beside the other.

Apache Spark provides the join() method to join DataFrames. Though Spark DataFrame join would probably occupy a complete book, we are going to limit its scope to the five most widely used join types: inner, outer, left, right, and anti joins.

The Spark DataFrame INNER join is the most popular. INNER join combines DataFrames including only the common elements to all DataFrames involved.

As usual, let's see how INNER join works with a practical example, and for that, the first step is to create a couple of DataFrames to work with. The first one contains the attributes of users from different nationalities including a field identifying the ISO 3166 country code of their country:

```
val dfUByL=spark.read.option("header", "true").csv("file:///Users/
aantolinez/Downloads/User_by_language.csv")
// Output
+---------+--------+------+--------+-----------+------+
|firstName|lastName|gender|language|ISO3166Code|salary|
+---------+--------+------+--------+-----------+------+
|  Liselda|   Rojas|Female| Spanish|        484| 62000|
| Leopoldo|   Galán|  Male| Spanish|        604| 47000|
|  William|   Adams|  Male| English|        826| 99000|
|    James|   Allen|  Male| English|        124| 55000|
|   Andrea|   López|Female| Spanish|        724| 95000|
+---------+--------+------+--------+-----------+------+
```

The second one includes the names of different countries together with their ISO 3166 country codes:

```
val dfCCodes=spark.read.option("header", "true").csv("file:///Users/
aantolinez/Downloads/ISO_3166_country_codes.csv")
// Output
+-----------+--------------+
| ISO3166Code|   CountryName|
+-----------+--------------+
|        484|        Mexico|
|        826|United Kingdom|
|        250|        France|
|        124|        Canada|
|        724|         Spain|
+-----------+--------------+
```

Let's now use an INNER join to join the preceding two DataFrames on the ISO3166Code column:

```
dfUByL.join(dfCCodes, dfUByL("ISO3166Code") === dfCCodes("ISO3166Code"),
"inner").show()
```

```
// Output
```

```
+---------+---------+------+--------+-----------+------+-----------+--------------+
|firstName| lastName|gender|language|ISO3166Code|salary|ISO3166Code|   CountryName|
+---------+---------+------+--------+-----------+------+-----------+--------------+
|  Liselda|    Rojas|Female| Spanish|        484| 62000|        484|        Mexico|
| Leopoldo|    Galán|  Male| Spanish|        604| 47000|        604|          Peru|
|  William|    Adams|  Male| English|        826| 99000|        826|United Kingdom|
|    James|    Allen|  Male| English|        124| 55000|        124|        Canada|
|   Andrea|    López|Female| Spanish|        724| 95000|        724|         Spain|
|   Sophia|Rochefort|Female|  French|        250| 49000|        250|        France|
|      Ben|   Müller|Female|  German|        276| 47000|        276|       Germany|
+---------+---------+------+--------+-----------+------+-----------+--------------+
```

Do you see something in the preceding output? Yes, the join column is duplicated. This situation will put you into trouble if you try to work with the final DataFrame, as duplicate columns will create ambiguity. Therefore, it is very likely you will receive a message similar to this one: "org.apache.spark.sql.AnalysisException: Reference 'ISO3166Code' is ambiguous, could be: ISO3166Code, ISO3166Code."

One way to avoid this kind of problem is using a temporary view and selecting just the fields you would like to have. Let's repeat the previous example, but this time using the Spark function createOrReplaceTempView() to create a temporary view:

```
// Create a temporary view
dfUByL.createOrReplaceTempView("UByL")
dfCCodes.createOrReplaceTempView("CCodes")
// Now you can run a SQL query as you would do in a RDBMS
val cleanDf=spark.sql("SELECT u.*, c.CountryName FROM UByL u INNER JOIN
CCodes c ON u.ISO3166Code == c.ISO3166Code")
cleanDf.show(5)
// Output
```

```
+---------+--------+------+--------+-----------+------+--------------+
|firstName|lastName|gender|language|ISO3166Code|salary|   CountryName|
+---------+--------+------+--------+-----------+------+--------------+
|  Liselda|   Rojas|Female| Spanish|        484| 62000|        Mexico|
| Leopoldo|   Galán|  Male| Spanish|        604| 47000|          Peru|
|  William|   Adams|  Male| English|        826| 99000|United Kingdom|
```

```
|    James|    Allen|  Male|  English|       124|  55000|            Canada|
|   Andrea|    López|Female|  Spanish|       724|  95000|             Spain|
+---------+---------+------+---------+----------+------+--------------+
```

On the other hand, a `fullouter`, `outer`, or `full` join collects all rows from both DataFrames and adds a null value for those records that do not have a match in both DataFrames. Once more, we are going to show you how to use this kind of join with a practical example using the previous DataFrames. Please notice that the next three code snippets return exactly the same outcome:

```
// Fullouter, Full and Outer join
dfUByL.join(dfCCodes, dfUByL("ISO3166Code") === dfCCodes("ISO3166Code"),
"fullouter").show()

dfUByL.join(dfCCodes, dfUByL("ISO3166Code") === dfCCodes("ISO3166Code"),
"full").show()

dfUByL.join(dfCCodes, dfUByL("ISO3166Code") === dfCCodes("ISO3166Code"),
"outer").show()
// Output
```

firstName	lastName	gender	language	ISO3166Code	salary	ISO3166Code	CountryName
James	Allen	Male	English	124	55000	124	Canada
Agnete	Jensen	Female	Danish	208	80000	null	null
Sophia	Rochefort	Female	French	250	49000	250	France
Ben	Müller	Female	German	276	47000	276	Germany
Amelie	Hoffmann	Female	German	40	45000	null	null
Liselda	Rojas	Female	Spanish	484	62000	484	Mexico
Leopoldo	Galán	Male	Spanish	604	47000	604	Peru
Andrea	López	Female	Spanish	724	95000	724	Spain
William	Adams	Male	English	826	99000	826	United Kingdom

The Spark left outer join collects all the elements from the left DataFrame and only those from the right one that have a matching on the left DataFrame. If there is no matching element on the left DataFrame, no join takes place.

Once again, the next code snippets will give you the same result:

```
dfUByL.join(dfCCodes, dfUByL("ISO3166Code") === dfCCodes("ISO3166Code"),
"left").show()
```

```
dfUByL.join(dfCCodes, dfUByL("ISO3166Code") === dfCCodes("ISO3166Code"),
"leftouter").show()
// Output
```

```
+---------+---------+------+--------+-----------+------+-----------+--------------+
|firstName| lastName|gender|language|ISO3166Code|salary|ISO3166Code|   CountryName|
+---------+---------+------+--------+-----------+------+-----------+--------------+
|  Liselda|    Rojas|Female| Spanish|        484| 62000|        484|        Mexico|
| Leopoldo|    Galán|  Male| Spanish|        604| 47000|        604|          Peru|
|  William|    Adams|  Male| English|        826| 99000|        826|United Kingdom|
|    James|    Allen|  Male| English|        124| 55000|        124|        Canada|
|   Andrea|    López|Female| Spanish|        724| 95000|        724|         Spain|
|   Sophia|Rochefort|Female|  French|        250| 49000|        250|        France|
|      Ben|   Müller|Female|  German|        276| 47000|        276|       Germany|
|   Agnete|   Jensen|Female|  Danish|        208| 80000|       null|          null|
|   Amelie| Hoffmann|Female|  German|         40| 45000|       null|          null|
+---------+---------+------+--------+-----------+------+-----------+--------------+
```

The Spark right outer or right join performs the left join symmetrical operation. In this case, all the elements from the right DataFrame are collected, and a null value is added where no matching is found on the left one. Next is an example, and again, both lines of code produce the same outcome:

```
dfUByL.join(dfCCodes, dfUByL("ISO3166Code") === dfCCodes("ISO3166Code"),
"right").show()
```

```
dfUByL.join(dfCCodes, dfUByL("ISO3166Code") === dfCCodes("ISO3166Code"),
"rightouter").show()
// Output
```

```
+---------+---------+------+--------+-----------+------+-----------+--------------+
|firstName| lastName|gender|language|ISO3166Code|salary|ISO3166Code|   CountryName|
+---------+---------+------+--------+-----------+------+-----------+--------------+
|  Liselda|    Rojas|Female| Spanish|        484| 62000|        484|        Mexico|
|  William|    Adams|  Male| English|        826| 99000|        826|United Kingdom|
```

```
|   Sophia|Rochefort|Female|   French|      250| 49000|       250|      France|
|    James|   Allen|  Male| English|      124| 55000|       124|      Canada|
|   Andrea|   López|Female| Spanish|      724| 95000|       724|       Spain|
| Leopoldo|   Galán|  Male| Spanish|      604| 47000|       604|        Peru|
|      Ben|  Müller|Female|  German|      276| 47000|       276|     Germany|
+---------+--------+------+--------+----------+------+----------+------------+
```

Finally, the Spark anti join returns rows from the first DataFrame not having matches in the second one. Here is one more example:

```
dfUByL.join(dfCCodes, dfUByL("ISO3166Code") === dfCCodes("ISO3166Code"),
"anti").show()
// Output
+---------+--------+------+--------+-----------+------+
|firstName|lastName|gender|language|ISO3166Code|salary|
+---------+--------+------+--------+-----------+------+
|   Agnete|  Jensen|Female|  Danish|        208| 80000|
|   Amelie|Hoffmann|Female|  German|         40| 45000|
+---------+--------+------+--------+-----------+------+
```

Summarizing, in this section you have seen how to use the most typical Spark joins. However, you have to bear something important in mind. Joins are wide Spark transformations; therefore, they imply data shuffling across the nodes. Hence, performance can be seriously affected if you use them without caution.

4.3 Spark Cache and Persist of Data

We have already mentioned in this book that one of Spark's competitive advantages is its data partitioning capability across multiple executors. Splitting large volumes of information across the network poses important challenges such as bandwidth saturation and network latency.

While using Spark you might need to use a dataset many times over a period of time; therefore, fetching the same dataset once and again to the executors could be inefficient. To overcome this obstacle, Spark provides two API calls called cache() and persist() to store locally in the executors as many of the partitions as the memory permits. Therefore, cache() and persist() are Spark methods intended for iterative and interactive application performance improvement.

Spark `cache()` and `persist()` are equivalent. In fact when `persist()` is called without arguments, it internally calls `cache()`. However, `persist()` with the `StorageLevel` argument offers additional storage optimization capabilities such as whether data should be stored in memory or on disk and/or in a serialized or unserialized way.

Now we are going to see with a practical example the impact the use of `cache()` can have in Spark operation performance:

```
// We create a 10^9 DataFrame
val dfcache = spark.range(1, 1000000000).toDF("base").withColumn("square",
$"base" * $"base")
dfcache.cache() // Cache is called to create a data copy in memory
spark.time(dfcache.count()) // Cache is materialized only first execute
an action
```

```
Time taken: 57407 ms
Out[96]: res81: Long = 1000000000
spark.time(dfcache.count())  // This time count() takes advantage of the
cached data
```

```
Time taken: 2358 ms
Out[98]: res83: Long = 1000000000
```

If you look attentively at the preceding example, you can see that cache() is lazily evaluated; it means that it is not materialized when it is called, but the first time it is invoked. Thus, the first call to `dfcache.count()` does not take advantage of the cached data; only the second call can profit from it. As you can see, the second `dfcache.count()` executes 24.34 times faster.

Is it worth noticing that `cache()` persists in memory the partitioned data in unserialized format. The cached data is localized in the node memory processing the corresponding partition; therefore, if that node is lost in the next invocation to that information, it would have to be recovered from the source. As the data will not be serialized, it would take longer and perhaps produce network bottleneck.

To overcome the `cache()` restrictions, the `DataFrame.persist()` method was introduced and accepts numerous types of storage levels via the `'storageLevel' [ = ]` value key and value pair.

The most typical valid options for storageLevel are

- NONE: With no options, `persist()` calls `cache()` under the hood.

- DISK_ONLY: Data is stored on disk rather than in RAM. Since you are persisting on disk, it is serialized in nature.

- MEMORY_ONLY: Stores data in RAM as deserialized Java objects. Full data cache is not guaranteed as it cannot be fully accommodated into memory and has no replication.

- MEMORY_ONLY_SER: Stores data as serialized Java objects. Generally more space-efficient than deserialized objects, but more read CPU-intensive.

- MEMORY_AND_DISK: Stores data as deserialized Java objects. If the whole data does not fit in memory, store partitions not fitting in RAM to disk.

- OFF_HEAP: It is an experimental storage level similar to MEMORY_ONLY_SER, but storing the data in off-heap memory if off-heap memory is enabled.

- MEMORY_AND_DISK_SER: Option Akin to MEMORY_ONLY_SER; however, partitions not fitting in memory are streamed to disk.

- DISK_ONLY_2, DISK_ONLY_3, MEMORY_ONLY_2, MEMORY_AND_DISK_2, MEMORY_AND_DISK_SER_2, MEMORY_ONLY_SER_2: The same as the parent levels adding partition replication to two cluster nodes.

Next, we are going to show you how to use persist() with a practical example:

```
import org.apache.spark.storage.StorageLevel
val dfpersist = spark.range(1, 1000000000).toDF("base").
withColumn("square", $"base" * $"base")
dfpersist.persist(StorageLevel.DISK_ONLY) // Serialize and cache
data on disk
spark.time(dfpersist.count()) // Materialize the cache

Time taken: 54762 ms
Out[107]: res90: Long = 999999999
```

```
spark.time(dfpersist.count()) // Taking advantage of cached data
```

```
Time taken: 2045 ms
Out[108]: res91: Long = 999999999
```

Once again, you can see in the preceding example the operation performed on persisted data is 26.8 times faster.

Additionally, tables and views derived from DataFrames can also be cached. Let's see again how it can be used with a practical example:

```
val dfWithQuery = spark.range(1, 1000000000).toDF("base").
withColumn("square", $"base" * $"base")
dfQuery.createOrReplaceTempView("TableWithQuery")
spark.sql("CACHE TABLE TableWithQuery")
spark.time(spark.sql("SELECT count(*) FROM TableWithQuery")).show()
```

```
Time taken: 2 ms
+---------+
| count(1)|
+---------+
|999999999|
+---------+
```

```
// Using already cached data
spark.time(spark.sql("SELECT count(*) FROM TableWithQuery")).show()
```

```
Time taken: 1 ms
+---------+
| count(1)|
+---------+
|999999999|
+---------+
```

Unpersisting Cached Data

In a similar way, data not in use can be unpersisted to release space using unpersist(), though Spark monitors the use your applications make of cache() and persist() and releases persisted data when it is not used. Spark also uses the Least Recently Used

(LRU) Page Replacement algorithm. Whenever a new block of data is addressed and not present in memory, Spark replaces one of the existing blocks with a newly created one.

Let's see how to use unpersist() taking advantage of the previous example:

```
val dfUnpersist = dfWithQuery.unpersist()

Out[5]: dfUnpersist: dfWithQuery.type = [base: bigint, square: bigint]

spark.time(dfUnpersist.count())

Time taken: 99 ms
Out[6]:    res2: Long = 9999999
```

Summarizing, in this section you have seen the important performance improvements you can get when using Spark caching techniques. In general, you should cache your data when you expect to use it several times during a job, and the storage level to use depends on the use case at hand. MEMORY_ONLY is CPU-efficient and performance-optimized. MEMORY_ONLY_SER used together with a serialization framework like Kyro is store-optimized and specially indicated when you have a DataFrame with many elements. Storage types involving replication, like MEMORY_ONLY_2, MEMORY_ONLY_2, and so on, are indicated if full fast recovery is required.

4.4 Summary

In this chapter we have explained what is called the Apache Spark high-level API. We have reviewed the concept of DataFrames and the DataFrame attributes. We have talked about the different methods available to create Spark DataFrames. Next, we explained how DataFrames can be used to manipulate and analyze information. Finally, we went through the options available to speed up data processing by caching data in memory. In the next chapter, we are going to study another Spark high-level data structure named datasets.

CHAPTER 5

Spark Dataset API and Adaptive Query Execution

In the Spark ecosystem, DataFrames and datasets are higher-level APIs that use Spark RDDs under the hood. Spark developers mainly use DataFrames and datasets because these data structures are the ones more efficiently using Spark storage and query optimizers, hence achieving the best data processing performance. Therefore, DataFrames and datasets are the best Spark tools in getting the best performance to handle structured data. Spark DataFrames and datasets also allow technicians with a RDBMS and SQL background to take advantage of Spark capabilities quicker.

5.1 What Are Spark Datasets?

According to the official Spark dataset documentation, datasets are *"a strongly typed collection of domain-specific objects that can be parallelized employing functional or relational operations."* Datasets were introduced in Spark 1.6 to overcome some dataframe limitations, and in Spark 2.0 both high-level APIs (the DataFrame and the Dataset) were merged into a single one, the Dataset API. Therefore, DataFrames can be thought of as datasets of type row or Dataset[Row].

Datasets combine RDD features like compile-time type safety and the capacity to use lambda functions with dataframe features including SQL automatic optimization. Datasets also incorporate compile-time safety, a feature only implemented in compiled languages like Java or Scala but not available in interpreted languages like PySpark or SparkR. That is why datasets are only available for Java and Scala.

© Alfonso Antolínez García 2023
A. Antolínez García, *Hands-on Guide to Apache Spark 3*, https://doi.org/10.1007/978-1-4842-9380-5_5

5.2 Methods for Creating Spark Datasets

A dataset can be created following four different ways:

- Can be created from a sequence of elements

- Can be created from a sequence of case classes

- Can be created from a RDD

- Can be created from a DataFrame

A dataset can be created from a sequence of elements using the toDS() method as in the following example:

```
val sequencia = Seq(0, 1, 2, 3, 4)
val sequenciaToDS = sequencia.toDS()
sequenciaToDS.show()
// Output
+-----+
|value|
+-----+
|    0|
|    1|
|    2|
|    3|
|    4|
+-----+
```

To create a dataset from a sequence of case classes using again the toDS() method, we first need a case class. Let's see how it works with a practical example. First, we create a Scala case class named Personas:

```
case class Personas(Nombre: String, Primer_Apellido: String,
Segundo_Apellido: String, Edad: Int, Sexo:String)
```

Then we can create a sequence of data matching the Personas case class schema. In this case we are going to use the data of some of the most famous Spanish writers of all times:

```
val personasSeq = Seq(
Personas("Miguel","de Cervantes","Saavedra",50,"M"),
Personas("Fancisco","Quevedo","Santibáñez Villegas",55,"M"),
Personas("Luis","de Góngora","y Argote",65,"M"))
```

After that we can obtain a dataset by applying the toDS() method to the personasSeq sequence:

```
val personasDs = personasSeq.toDS()
personasDs.show()
// Output
+--------+---------------+-------------------+----+----+
|  Nombre|Primer_Apellido|  Segundo_Apellido|Edad|Sexo|
+--------+---------------+-------------------+----+----+
|  Miguel|   de Cervantes|          Saavedra|  50|   M|
|Fancisco|        Quevedo|Santibáñez Villegas|  55|   M|
|    Luis|     de Góngora|          y Argote|  65|   M|
+--------+---------------+-------------------+----+----+
```

Another way of creating a dataset is from a RDD. Again we are going to show how it works with an example:

```
val myRdd = spark.sparkContext.parallelize(Seq(("Miguel de Cervantes",
1547),("Lope de Vega", 1562),("Fernando de Rojas",1470)))
val rddToDs = myRdd.toDS
.withColumnRenamed("_1","Nombre")
.withColumnRenamed("_2","Nacimiento")
rddToDs.show()
// Output
+-------------------+----------+
|             Nombre|Nacimiento|
+-------------------+----------+
|Miguel de Cervantes|      1547|
|       Lope de Vega|      1562|
|  Fernando de Rojas|      1470|
+-------------------+----------+
```

The fourth way of creating a dataset is from a DataFrame. In this case we are going to use an external file to create a DataFrame and after that transform it into a dataset, as you can see in the following code snippet:

```
case class Personas(Nombre: String, Primer_Apellido: String,
Segundo_Apellido: String, Edad: Int, Sexo:String)
val personas = spark.read.format("csv")
      .option("header", "true")
      .option("inferSchema", "true")
      .load("personas.csv")
      .as[Personas]
personas.show()
// Output
+--------+----------------+-------------------+----+----+
|  Nombre|  Primer_Apellido|  Segundo_Apellido|Edad|Sexo|
+--------+----------------+-------------------+----+----+
|  Miguel|     de Cervantes|           Saavedra|  50|   M|
|Fancisco|          Quevedo|Santibáñez Villegas|  55|   M|
|    Luis|       de Góngora|           y Argote|  65|   M|
|  Teresa|Sánchez de Cepeda|         y Ahumada|  70|   F|
+--------+----------------+-------------------+----+----+
```

5.3 Adaptive Query Execution

With each release, Apache Spark has introduced new methods of improving performance of data querying. Therefore, with Spark 1.x, the Catalyst Optimizer and Tungsten Execution Engine were introduced; Spark 2.x incorporated the Cost-Based Optimizer; and finally, with Spark 3.0 the new Adaptive Query Execution (AQE) has been added.

The Adaptive Query Execution (AQE) is, without a doubt, one of the most important features introduced with Apache Spark 3.0. Before Spark 3.0, query execution plans were monolithic; it means that before executing a query, Spark established an execution plan and once the job execution began this plan was strictly followed independently of the statistics and metrics collected at each job stage. In Figure 5-1 the process followed by Spark 2.x to create and execute a query using the cost-based optimization framework is depicted.

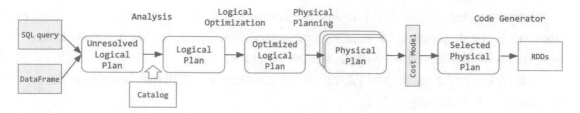

Figure 5-1. *Spark 2.x Cost-Based Optimizer (Source: Databricks)*

Cost-based optimization tries to improve the quality of executing a SQL statement by generating multiple execution plans employing execution rules and at the same time, calculating the computing cost of each query. Cost-based optimization techniques can improve decisions such as selecting the most efficient join type, selecting the correct build side in a hash join, and so on.

The introduction of the Adaptive Query Execution permits the Spark SQL engine a continuous update of the execution plan while running based on the statistics collected at runtime. The AQE framework includes the following three major features:

- Data-dependent adaptive determination of the shuffle partition number

- Runtime replanning of join strategies according to the most accurate join relation size

- Optimization of unevenly distributed data joins at execution time

The Spark 3 Adaptive Query Execution (AQE) framework introduces the concepts of *materialization points* and *query stages*. The AQE succinctly works as follows: Normally, Spark processes run pipelined and in parallel; however, actions triggering shuffling and broadcasting operations break that pipelining as each query stage must materialize its intermediate results, and all running parallel processes have to materialize their intermediate stages before the workflow can continue. These temporal interruptions offer an occasion for query plan reoptimization.

When a query is released, the Adaptive Query Execution framework launches all the so-called leaf stages or stages that do not depend on any other. As soon as these stages finish their individual materializations, the AQE labels them as completed in the physical query plan and subsequently updates the logical query plan metrics with the fresh statistics collected from finished stages at runtime. Founded on updated information, the AQE reruns the optimizer and the physical planner.

With the now retrofitted and hopefully improved execution plan, the AQE framework runs the query stages under those already materialized, repeating this run-reevaluate-execute cycle until the entire query has been executed.

With the new Adaptive Query Execution framework, the SQL plan workflow looks like the one shown in Figure 5-2.

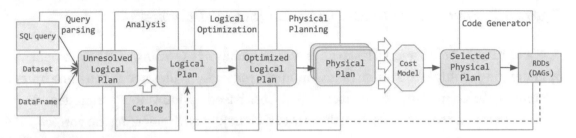

Figure 5-2. *Spark 3 Adaptive Query Execution framework SQL plan*

5.4 Data-Dependent Adaptive Determination of the Shuffle Partition Number

According to the Adaptive Query Execution official documentation, by default Apache Spark sets to 200 the number of partitions to use for data shuffling in join and aggregation operations. This base parameter is not necessarily the option at any time because shuffle operations have a very important impact on Spark performance when we are dealing with very large datasets as these operations require the rearrangement of the information across the nodes of the cluster and movement of data through the network.

In order to minimize the impact of data transfer, the allocation of the correct number of partitions is key. On the other hand, the number of partitions is heavily dependent on data volume, which in turn varies from stage to stage and operation to operation.

- If the number of allocated partitions is smaller than necessary, the size of the information per partition could be too big to fit in memory. Hence, writing it down to disk could be necessary, jeopardizing performance.

- In case the number of partitions is greater than required, it can introduce additional bottlenecks:

    - Firstly, the size of every partition could be very small, requiring a large number of I/O operations to fetch the data blocks involved in the shuffle operations.

    - Secondly, too many operations can increase the number of jobs Spark needs to handle, therefore creating additional overhead.

To tackle these problems, you can set your best estimation of the number of partitions and let Spark dynamically adapt this number according to the statistics collected between stages. Possible strategies to choose the best numPartitions can be

- Based on computer capacity available

- Based on the grounds of the amount of data to be processed

A graphical example of how the Adaptive Query Execution works is illustrated in Figure 5-3.

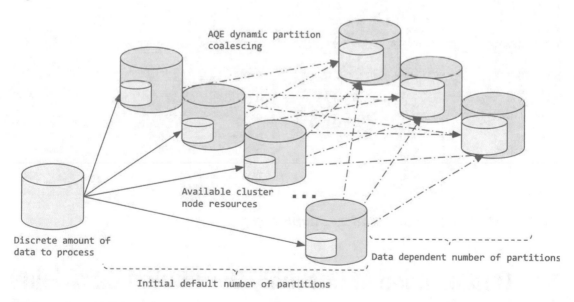

Figure 5-3. *Adaptive Query Execution (AQE)*

5.5 Runtime Replanning of Join Strategies

With the AQE enabled, when we want to perform join operations on tables, Spark can determine the optimal join algorithm to use at runtime. Spark supports several join strategies. Among them the broadcast hash join is usually the most performant one when one of the sides of the join can fit in memory and it is smaller than the broadcast threshold.

With the new AQE enabled, Spark can replan at runtime the join strategy according to the information collected. As you can see in Figure 5-4, Spark develops an initial join strategy based on the information available at the very beginning. After first computations and based on data volumes and distribution, Spark is capable of adapting the join strategy from an initial sort merge join to a broadcast hash join more suitable for the situation.

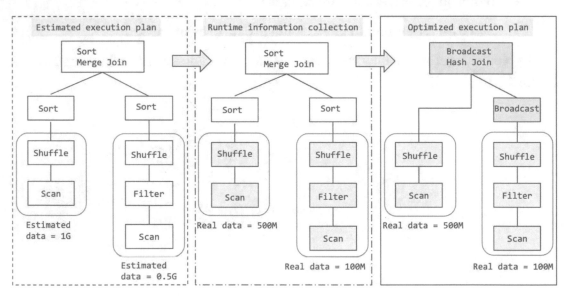

Figure 5-4. *Dynamic join strategy adaptation*

5.6 Optimization of Unevenly Distributed Data Joins

A skewed dataset is characterized by a plot frequency distribution not perfectly symmetrical but skewed to the left or right side of the graph. Said another way, data skew is associated with an uneven or nonuniform distribution of data among different partitions in a cluster. In Figure 5-5 you can see a graphical representation of a real positively skewed dataset.

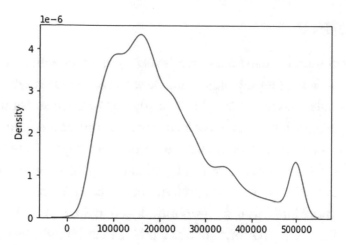

Figure 5-5. *Positive skew*

In real-life scenarios, you will often have to tackle the problem of non-ideal data distributions. Severe data skewness can severely jeopardize Spark join performance because for join operations, Spark has to place records of each key in its particular partition. Therefore, if you want to join two dataframes by a specific key or column and one of the keys has many more records than the others, its corresponding partition will become much bigger than the others (or skewed); therefore, the time taken to process that partition will be comparatively longer than the time consumed by others, consequently causing job bottlenecks, poor CPU utilization, and/or out-of-memory problems.

The AQE automatically detects data skewness from shuffle statistics and divides the bigger partitions into smaller ones that will be joined locally with their corresponding counterparts.

For Spark, to take advantage of skew join optimization, both options "spark.sql.adaptive.enabled" and "spark.sql.adaptive.skewJoin.enabled" have to be set to true.

5.7 Enabling the Adaptive Query Execution (AQE)

The Adaptive Query Execution is disabled by default in Spark 3.0. In order to enable it, you must set the `spark.sql.adaptive.enabled` configuration property to true. However, advantages of the AQE can only be applied to not streaming queries or when they include operations entailing data exchange such as joins, aggregates, or window operators.

5.8 Summary

Datasets are part of the so-called Spark high-level API together with DataFrames. However, unlike the latter, they are only available with compiled programming languages, such as Java and Scala. This attribute is both an advantage and disadvantage as those programming languages' learning curves have a more pronounced gradient than that of Python, for example; hence, they are less commonly employed. Datasets also provide security improvements as they are strongly typed data structures. After introducing the concept of datasets, we focused on the Spark Adaptive Query Execution (AQE) as it is one of the newest and more interesting features introduced in Spark 3.0. The AQE permits the improvement of Spark query performance as it is able to automatically adapt query plans based on statistical data collected at runtime. In the coming chapters, we have to switch to another important Spark feature, which is data streaming.

CHAPTER 6

Introduction to Apache Spark Streaming

Spark Streaming was added to Apache Spark in 2013 as a scalable, fault-tolerant, real-time streaming processing extension of the core Spark API. Spark Streaming natively supports both batch and streaming workloads and uses micro-batching to ingest and process streams of data passing the results to permanent storage or a dashboard for online data analysis.

Spark Streaming incorporates a large ecosystem of data sources and data sinks. Among the former we can include Apache Kafka, Apache Flume, Amazon Kinesis, and TCP sockets as data streamers, and among the latter, we can include most of the RDBMS and NoSQL databases such as MemSQL, PostgreSQL, and MySQL, including of course file storage format such as Parquet or CSV.

6.1 Real-Time Analytics of Bound and Unbound Data

Continuous streaming and real-time analytics are changing the way we consume information nowadays. In fact, in the real world most of the information we produce and consume is based on unbound or unconfined data. For example, our brain is processing data and taking decisions as information is coming through in very definite time windows.

In mathematics, *unbounded* means that a function is not confined or bounded. Therefore, an unbounded set is a set that has no finite upper or lower limits. Examples of unbounded sets could be $(-\infty,+\infty)$, $(5,+\infty)$, etc. On the other hand, a set that has finite values of upper and lower bounds is said to be bounded. That is why you would probably hear that unbounded data is (theoretically) infinite.

© Alfonso Antolínez García 2023

A. Antolínez García, *Hands-on Guide to Apache Spark 3*, https://doi.org/10.1007/978-1-4842-9380-5_6

Data analytics is divided into batch and streaming processing. Traditionally, batch data is associated with what is commonly known as bound data.

Bound data is steady, unchanging, with a known size, and within a defined timeframe. Examples of this confined data are last year or last quarter sales data analysis, historical stock market trades, and so on. On the other hand, unbound or unconfined data is in motion (is not finished) and very often not in a perfect or expected sequence.

Spark Streaming works as follows: by receiving continuous data streams and dividing them into micro-batches called Discretized Streams or DStreams, which are passed to the Spark engine that generates batches of results.

A graphical depiction of continuous stream processing of batch streams of data to provide real-time data analytics is shown in Figure 6-1.

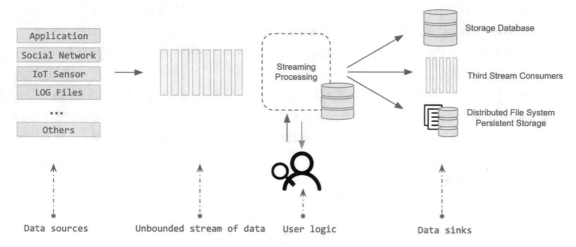

Figure 6-1. *Continuous stream processing enabling real-time analytics*

6.2 Challenges of Stream Processing

As we analyze data using finite resources (finite number of CPUs, RAM, storage hardware, etc.), we cannot expect to be able to accommodate it in finite resources. Instead, we have to process it as a sequence of events received over a period of time with limited computational resources.

While processing data streams, time introduces important challenges because there is a time difference between the moment the information is produced and the moment it is processed and analyzed.

There are three important concepts when we deal with streaming data processing:

- *Event time*: It refers to the moment when the event was produced on its producing device. This time reference can be established by the event emitter (typically) or by the event receiver; either way, you have to understand the additional complexity brought by operating in different timezones, for example.

- *Ingestion time*: It is the time when an event flows into the streaming process.

- *Data processing time*: In this case the frame of reference is the computer in which data is being processed.

6.3 The Uncertainty Component of Data Streams

Another important aspect to consider when dealing with data streams is the uncertainty over data throughput. In general, no assumptions can be made about the data cadence arriving to the system; therefore, it is not possible to precisely foresee either the future hardware resources needed or the order in the sequence of events.

Information coming from stream sources such as sensors, social networks, and so on can suffer delays and interruptions due to numerous circumstances. In these particular situations, one of the following can happen: Either the information is piled up, and it is released as a kind of data avalanche as soon as the connection is restored. Or the data is lost; hence, some gap in the sequence of events will appear. When something like the preceding scenarios takes place, the system can be temporarily saturated by the oversupply of data, causing processing delays or failures.

6.4 Apache Spark Streaming's Execution Model

From inception, Apache Spark was conceived as an unified engine for batch and streaming data processing aiming to provide some major features such as

- A rapid service restoration from failures

- Active identification of slow-running tasks, a.k.a. stragglers, and actively dealing with them

- Improved load evenness and computation resource utilization

- The capacity of consuming data streams from static datasets and interactive SQL queries

- Built-in integration with other Spark modules like SQL, MLlib, etc.

Among the above-mentioned features, Spark's capability to tackle slow-running tasks or stragglers is worth mentioning. A straggler refers to a task within a stage that takes exceptionally longer time to execute than other ones belonging to the same stage. In terms of performance, it is always a good practice to keep an eye on stragglers as they can frequently appear in a stage or disseminated across multiple stages.

The appearance of slow-running tasks even in a single stage can considerably delay Spark jobs and produce a cascade effect damaging the overall performance of your application.

6.5 Stream Processing Architectures

In this section we introduce two of the most typical real-time data processing technological approaches, the so-called Lambda and Kappa architectures. Setting up a proper and economically sustainable real-time processing architecture that best fits our business needs is not something trivial.

Real-time data processing architectures are technological systems conceived to efficiently manage the full real-time data life cycle: data intake, data processing, and finally storage of huge amounts of data. They play a pivotal role in allowing modern businesses to gain valuable insights from the information available. These architectures together with data analytics techniques are the foundations of the modern data-driven organizations that take advantage of data analysis to improve decision-making, achieve competitive advantage and operational efficiency, and ultimately drive growth.

The Lambda and Kappa architectures are designed to cope with both batch processing and real-time data processing.

The Lambda Architecture

The Lambda architecture was developed by Nathan Marz, the creator of Apache Storm, in 2011 as a scalable, fault-tolerant, and flexible architecture for massive real-time data processing.

The main characteristic of the Lambda architecture is that it incorporates two separate lanes to process different kinds of workloads as can be seen in Figure 6-2. The batch lane is intended to process large amounts of data and store the results in a centralized warehouse or distributed file system such as Hadoop.

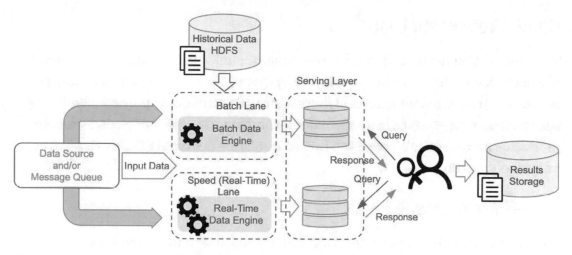

Figure 6-2. *Lambda architecture for real-time stream and batch data processing*

The real-time lane takes care of data as it arrives to the system, and as the batch lane, it stores the result in a distributed data warehouse system.

The Lambda architecture has shown to satisfy many business use cases, and it is currently in use by important corporations like Yahoo and Netflix.

The Lambda architecture is integrated by three main layers or lanes:

- Batch processing layer

- Real-time or speed layer

- Serving layer

In addition to these three main data processing lanes, some authors would add a pre-layer for data intake:

- Data ingestion layer

Let's succinctly review all of them.

Data Ingestion Layer

This lane is responsible for integrating on the fly the raw data coming from several data sources. This information is supplied to the batch and speed lanes simultaneously.

Batch Processing Layer

This lane is intended to efficiently process large amounts of information and to provide a holistic view of the data. The batch layer is responsible for (a) integrating historical consolidated data into the analytical process and (b) reprocessing previous results such as retraining machine learning models. This layer oversees the full dataset, hence producing more precise outcomes; however, results are delivered offline as it takes a longer computation time.

Real-Time or Speed Layer

This layer is intended for providing a low-latency and almost up-to-the-minute vision of huge volumes of data streams complementing the batch layer with incremental outcomes. Thanks to these incremental results, computation time is decreased. This layer is responsible for real-time data processing and stores the results obtained in a distributed storage (NoSQL databases or file systems).

Serving Layer

This layer merges the results from the batch and real-time layers and constitutes the way users use to interactivcly submit queries and receive the results online. This layer allows users to seamlessly interact with full data being processed independently of whether it is being processed on the batch or stream lane. This lane also provides the visualization layer with up-to-the-minute information.

Pros and Cons of the Lambda Architecture

The Lambda architecture meets many big data use cases, but at a high cost in terms of redundancy and complexity. Therefore, it has pros and cons.

Pros:

- *Scalability*: The Lambda architecture is suitable for horizontal scalability meeting big data requirements.

- Fault tolerance against hardware failures.

- Flexibility handling both batch and streaming workloads.

Cons:

- *Complexity*: Due to its distributed nature and different technologies being involved, this architecture is complex and redundant and could be difficult to tune and maintain.

- *Possible data discrepancies*: As data is processed through parallel lanes, processing failures can bring discrepant results from batch and stream lanes.

The Kappa Architecture

The Kappa architecture shown in Figure 6-3 was designed by Jay Kreps in 2014 to confront some of the problems identified in the Lambda architecture and to avoid maintaining two separate developments.

The Kappa architecture uses a single lane with two layers: stream processing and serving for both batch and stream data processing workloads, hence treating every data influx as streams of data. This architecture is simpler than Lambda as live stream data intake, processing, and storage is performed by the stream processing layer while still maintaining fast and efficient query capabilities.

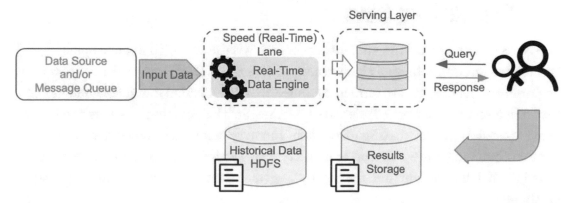

Figure 6-3. *Kappa architecture for real-time stream data processing*

The two layers of the Kappa architecture implement the following functionalities:

- *Stream processing*: This module is responsible for live data ingestion and permanent raw data storage.

- *Serving*: The serving layer is responsible for the provision of the necessary tools for data querying.

Pros and Cons of the Kappa Architecture

Here we introduce some of the pros and cons of the Kappa architecture.

Pros:

- Simplified design, implementation, debugging, and maintenance of the pipeline

- Facilitates pipeline migration and reorganization taking advantage of the single-lane pipeline

Cons:

- *Complexity*: Though it is simpler than the Lambda architecture, the overall infrastructure is still complex.

- *Scalability issues*: The use of a single streaming lane processing unbounded streams of data can provoke bottlenecks when it comes time to use the great volume of results processed.

6.6 Spark Streaming Architecture: Discretized Streams

The key Spark abstraction for streaming processing is the so-called Apache Spark Discretized Stream, commonly known as Spark DStream. DStreams, as we saw at the beginning of the chapter, are formed by a stream of data divided into micro-batches of less than a second. To create these micro-batches, Spark Streaming receivers can ingest data in parallel and store it in Spark workers' memory before releasing it to be processed. In micro-batching, the Spark worker nodes typically wait for a defined period of time—called the batch cycle—or until the batch size gets to the upper limit before executing the batch job.

For example, if a streaming job is configured with a batch cycle of 1 s and a batch size limit of 128 items read, even if the number of items read is less than 128 in a second, the job will start anyway. On the other hand, if the receivers sustain a high throughput and the number of items received is higher than 128 in less than 1 s, the job would start without waiting for the batch cycle to complete.

DStream is internally built on top of a series of unbounded Spark RDDs; therefore, transformations and actions executed on DStreams are in fact RDD operations. Each RDD contains only data from a certain slot of time as shown in Figure 6-4.

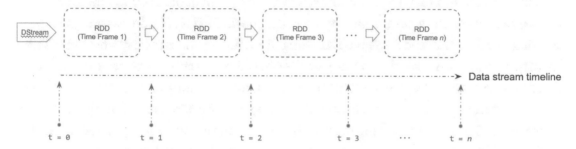

Figure 6-4. *Spark DStream as a succession of micro-batches (RDDs)*

The use of RDDs under the hood to manipulate data facilitates the use of a common API both for batch and streaming processing. At the same time, this architecture permits the use of any third-party library available to process Spark data streams.

Spark implements streaming load balancing and a faster fault recovery by dynamically assigning processing tasks to the workers available.

Thus, a Discretized Stream (DStream) is in fact a continuous sequence of RDDs of the same type simulating a continuous flow of data and bringing all the RDD advantages in terms of speed and safety to near-real-time stream data processing. However, the DStream API does not offer the complete set of transformations compared with the Apache Spark RDD (low-level API).

6.7 Spark Streaming Sources and Receivers

In Spark Streaming, input data streams are DStreams symbolizing the input point of incoming information and are the way to integrate data from external data sources.

Every input DStream (except file stream) is associated with a Receiver object, and it acts as a buffer between data sources and Spark workers' memory where data is piled up in micro-batches before it is processed.

There are two built-in streaming sources in Spark:

- Basic input sources

- Advanced input sources

Basic Input Sources

Basic sources are those natively built in the StreamingContext API, for example, TCP socket connections and file system sources. When a stream of data is implemented through a socket connection, Spark listens to the specified TCP/IP address and port creating input Discretized Streams (DStreams) from the text data received over that TCP socket connection. Socket streams are receiver-based connections; hence, they require the implementation of a receiver. Therefore, when implementing an input DStream based on a socket receiver, enough resources (CPU cores) must be allocated to receive data and to process it. Thus, when running receiver-based streams in local mode, either "local[*]" or "local[n]" (with n > number of receivers) has to be implemented as the master URL. For the same reason, if receiver-based streaming jobs are run in cluster mode, the number of processing threads available must be bigger than the number of receivers.

On the other hand, file streams do not require executing receivers; in consequence, no additional CPU resources are needed for file data ingestion. File sources are used to ingest data from files as they appear in a system folder, hence simulating a data stream. Input file formats supported are text, CSV, JSON, ORC, and Parquet.

Socket Connection Streaming Sources

A socket is the union of an IP address and a network port (<IP>:<port>). Sockets permit the connection between computers over the TCP/IP where each ending computer sets up a connection socket.

To show how Spark Streaming can be used to ingest data by listening to a socket connection, we are going to use an example based on a real data streaming use case; we are going to implement a basic near-real-time streaming Hospital Queue Management System with a CSV format as input as the ones you can see next:

```
1004,Tomás,30,DEndo,01-09-2022
1005,Lorena,50,DGineco,01-09-2022
1006,Pedro,10,DCardio,01-09-2022
```

1007,Ester,10,DCardio,01-09-2022
1008,Marina,10,DCardio,01-09-2022

1009,Julia,20,DNeuro,01-09-2022
1010,Javier,30,DEndo,01-09-2022
1011,Laura,50,DGineco,01-09-2022
1012,Nuria,10,DCardio,01-09-2022
1013,Helena,10,DCardio,01-09-2022
1014,Nati,10,DCardio,01-09-2022

Next, we show two options to see our program up and running. The first code (*socketTextStream.scala*) is shown next, and it is a Scala variant that can be compiled and executed in Spark using the *$SPARK_HOME/bin/spark-submit* command. It is out of the scope of this book to discuss how to compile and link Scala code, but it is recommended to use sbt[1] together with sbt-assembly[2] to create a so-called "*fat JAR*" file including all the necessary libraries, a.k.a. *dependencies*:

```scala
import org.apache.spark.sql.SparkSession
import org.apache.spark.streaming.{Seconds, StreamingContext}
import java.io.IOException

object socketTextStream {
  def main(args: Array[String]): Unit = {
    val host = "localhost"
    val port = 9999

    try {
    val spark: SparkSession = SparkSession.builder()
    .master("local[*]")
    .appName("Hand-On-Spark3_socketTextStream")
    .getOrCreate()

    spark.sparkContext.setLogLevel("ERROR")

    val sc = spark.sparkContext
    val ssc = new StreamingContext(sc, Seconds(5))
```

---

[1] www.scala-sbt.org/
[2] https://github.com/sbt/sbt-assembly

```
    val lines = ssc.socketTextStream(host, port)

    printf("\n Spark is listening on port 9999 and ready...\n")

    lines.print()
    ssc.start()
    ssc.awaitTermination()

    } catch {
    case e: java.net.ConnectException => println("Error establishing
    connection to " + host + ":" + port)
    case e: IOException => println("IOException occurred")
    case t: Throwable => println("Error receiving data", t)
    } finally {
    println("Finally block")
    }

  }
}
```

The next code snippet is a version of the preceding Hospital Queue Management System application that can be executed in Spark using a notebook application such as Jupyter,[3] Apache Zeppelin,[4] etc., which can be more convenient for learning purposes, especially if you are not familiar with Scala code compiler tools:

```
import org.apache.spark.SparkConf
import org.apache.spark.sql.SparkSession
import org.apache.spark.streaming.{Seconds, StreamingContext}
import java.io.IOException

val host = "localhost"
val port = 9999

try{
    val spark = SparkSession
    .builder()
    .master("local[*]")
```

---

[3] https://jupyter.org/
[4] https://zeppelin.apache.org/

```
    .appName("Hands-On_Spark3_socketTextStream")
    .getOrCreate()

    val sc = spark.sparkContext

    // Create the context with a 5 seconds batch size
    val ssc = new StreamingContext(sc, Seconds(5))

    val lines = ssc.socketTextStream(host, port)

    printf("\n Spark is listening on port 9999 and ready...\n")

    lines.print()
    ssc.start()
    ssc.awaitTermination()
}catch {
    case e: java.net.ConnectException => println("Error establishing
    connection to " + host + ":" + port)
    case e: IOException => println("IOException occurred")
    case t: Throwable => println("Error receiving data", t)
    } finally {
    println("Finally block")
    }
```

Pay attention to the local[*] option. In this case we have used "*"; thus, the program is going to use all the cores available. It is important to use more than one because the application must be able to run two tasks in parallel, listening to a TCP socket (localhost:9999) and, at the same time, processing the data and showing it on the console.

Running Socket Streaming Applications Locally

We are going to use a featured networking utility called Netcat[5] to set up a simple client/server streaming connection. Netcat (netcat, nc, ncat, etc., depending on the system) is available in Unix-like operating systems and uses the TCP/IP to read and write data through a network. In this book we use the Netcat OpenBSD version (nc).

---

[5] https://netcat.sourceforge.net/

The syntax for the nc command is

```
nc [<options>] <host> <port>
```

Netcat has several [<options>]; however, we are going to use only -l, which instructs nc to listen on a UDP or TCP <port>, and -k, which is used in listen mode to accept multiple connections. When <host> is omitted, nc listens to all the IP addresses bound to the <port> given.

To illustrate how the program works, we are going to take advantage of the nc utility introduced, to establish a streaming client/server connection between nc and our Spark application. In our case nc will act as a server (listens to a host:port), while our application will act as a client (connects to the nc server).

Whether you have built your JAR file from the previous code or are using the notebook version, running the application consists of a two-step process:

1. Open a terminal in your system and set up the server side of the client/server streaming connection by running the following code:

    ```
    nc -lk 9999
    ```

2. Depending on how you are running the application

    2.1. *Using a JAR file*: Open a second terminal and execute your application as shown in the following:

    ```
    $SPARK_HOME/bin/spark-submit --class org.
    apress.handsOnSpark3.socketTextStream --master
    "local[*]" /PATH/TO/socketTextStream/HandsOnSpark3-
    socketTextStream.jar
    ```

    2.2. *Using a notebook*: Just execute the code in your notebook.

    As soon as you see the message Spark is listening on port 9999 and ready... on your screen, you can go back to step 1 and type some of the CSV strings provided as examples, for instance:

    ```
    1009,Julia,20,DNeuro,01-09-2022
    1010,Javier,30,DEndo,01-09-2022
    1011,Laura,50,DGineco,01-09-2022
    1012,Nuria,10,DCardio,01-09-2022
    1013,Helena,10,DCardio,01-09-2022
    ```

```
1014,Nati,10,DCardio,01-09-2022

1004,Tomás,30,DEndo,01-09-2022
1005,Lorena,50,DGineco,01-09-2022
1006,Pedro,10,DCardio,01-09-2022
1007,Ester,10,DCardio,01-09-2022
1008,Marina,10,DCardio,01-09-2022
```

With a cadence of seconds, you will see an output like the following one coming up on your terminal or notebook:

```
Spark is listening on port 9999 and ready...
-------------------------------------------
Time: 1675025630000 ms
-------------------------------------------
1009,Julia,20,DNeuro,01-09-2022
1010,Javier,30,DEndo,01-09-2022
1011,Laura,50,DGineco,01-09-2022
1012,Nuria,10,DCardio,01-09-2022
1013,Helena,10,DCardio,01-09-2022
1014,Nati,10,DCardio,01-09-2022


-------------------------------------------
Time: 1675025635000 ms
-------------------------------------------
1004,Tomás,30,DEndo,01-09-2022
1005,Lorena,50,DGineco,01-09-2022
1006,Pedro,10,DCardio,01-09-2022
1007,Ester,10,DCardio,01-09-2022
1008,Marina,10,DCardio,01-09-2022
...
```

3. Application termination

   awaitTermination() waits for a user's termination signal. Thus, going to the terminal session started in step 1 and pressing Ctrl+C or SIGTERM, the streaming context will be stopped and your streaming application terminated.

However, this way of abruptly killing a streaming process is neither elegant nor convenient in most of the real streaming applications. The notion of unbounded data implies a continuous flow of information arriving to the system; thus, an abrupt interruption of the streaming process is bound to a loss of information in all likelihood in the majority of situations. To avoid data loss, a procedure is to halt a streaming application without suddenly killing it during the RDD processing; it is called a "graceful shutdown," and we are going to explain it later on in the "Spark Streaming Graceful Shutdown" section.

Improving Our Data Analytics with Spark Streaming Transformations

Though the number of transformations that can be applied to RDD DStreams is limited, the previous example can be tweaked to improve its functionality.

We can use the flatMap(*function()*) function, which takes function() as an argument and applies it to each element, returning a new RDD with 0 or more items.

Therefore, replacing the code line lines.print() in the previous code snippet with the following line

```
lines.flatMap(_.split(",")).print()
```

we get the following output:

```
-------------------------------------------
Time: 1675203480000 ms
-------------------------------------------
1009
Julia
20
DNeuro
01-09-2022
1010
Javier
30
DEndo
01-09-2022
...
```

```
-----------------------------------------------
Time: 1675203485000 ms
-----------------------------------------------
1005
Lorena
50
DGineco
01-09-2022
1006
Pedro
10
DCardio
01-09-2022
...
```

We can also introduce the count() function to count the number of lines in our stream. Thus, adding the count() function as you can see in the following line

```
lines.flatMap(_.split(",")).count().print()
```

and typing text lines from our example, we get an output similar to the following:

```
-----------------------------------------------
Time: 1675204465000 ms
-----------------------------------------------
75
-----------------------------------------------
Time: 1675204470000 ms
-----------------------------------------------
35
-----------------------------------------------
Time: 1675204470000 ms
-----------------------------------------------
260
```

We can also use countByValue() to count the number of occurrences of each word in the dataset, that is to say, the number of times each word occurs in the stream.

To achieve that goal, all we have to do is transform the

```
lines.flatMap(_.split(",")).print()
```

line of code into the following one:

```
lines.countByValue().print()
```

Running again the code and copying and pasting some of the lines provided as examples, you could see an output similar to the following:

```
Spark is listening on port 9999 and ready...
-------------------------------------------
Time: 1675236625000 ms
-------------------------------------------
(1013,Helena,10,DCardio,01-09-2022,1)
(1007,Ester,10,DCardio,01-09-2022,4)
(1010,Javier,30,DEndo,01-09-2022,1)
(1011,Laura,50,DGineco,01-09-2022,1)
(1005,Lorena,50,DGineco,01-09-2022,3)
(1014,Nati,10,DCardio,01-09-2022,1)
(1004,Tomás,30,DEndo,01-09-2022,1)
(1008,Marina,10,DCardio,01-09-2022,4)
(1012,Nuria,10,DCardio,01-09-2022,1)
(1006,Pedro,10,DCardio,01-09-2022,4)

-------------------------------------------
Time: 1675236630000 ms
-------------------------------------------
(1007,Ester,10,DCardio,01-09-2022,1)
(1005,Lorena,50,DGineco,01-09-202,1)
(1005,Lorena,50,DGineco,01-09-2022,1)
(1014,Nati,10,DCardio,01-09-2022,1)
(1004,Tomás,30,DEndo,01-09-2022,1)
(1008,Marina,10,DCardio,01-09-2022,1)
(1006,Pedro,10,DCardio,01-09-2022,1)
```

```
--------------------------------------------
Time: 1675236640000 ms
--------------------------------------------
(1007,Ester,10,DCardio,01-09-2022,1)
(1005,Lorena,50,DGineco,01-09-2022,1)
(1008,Marina,10,DCardio,01-09-2022,1)
(1006,Pedro,10,DCardio,01-09-2022,1)
```

We can improve our example even more and achieve the same result linking or piping the previous code line with the flatMap() function we saw before:

```
lines.flatMap(_.split(",")).countByValue().print()
```

As you previously did, run the code again, copy and paste the example lines in your terminal, and you will see again outcome similar to the next one:

```
Spark is listening on port 9999 and ready...
--------------------------------------------
Time: 1675236825000 ms
--------------------------------------------
(01-09-202,1)
(01-09-2022,20)
(1007,5)
(1008,5)
(50,4)
(DCardio,16)
(Lorena,4)
(Tomás,1)
(Marina,5)
(Ester,5)
...
--------------------------------------------
Time: 1675236830000 ms
--------------------------------------------
(01-09-202,1)
(01-09-2022,13)
(1007,3)
```

```
(1008,3)
(50,3)
(DCardio,10)
(Lorena,3)
(Tomás,1)
(Marina,3)
(Ester,3)
...
-------------------------------------------
Time: 1675236835000 ms
-------------------------------------------
(01-09-202,1)
(01-09-2022,6)
(1007,2)
(1008,1)
(50,2)
(DCardio,5)
(Lorena,2)
(Marina,1)
(Ester,2)
(DGineco,2)
...
```

This time the output is more informative, as you can see the Department of Cardiology registrations are piling up. That information could be used to, for example, trigger an alarm when the number of appointments approaches or crosses the threshold of maximum capacity.

We could have gotten the same result by using the reduceByKey() function. This function works on RDDs (key/value pairs) and is used to merge the values of each key using a provided reduce function (_ + _ in our example).

To do that, just replace the following line of code

```
lines.countByValue().print()
```

with the next one:

```
val words = lines.flatMap(_.split(",")).map(x => (x, 1)).reduceByKey(_+_)
words.print()
```

Repeating the process of copying and pasting example lines to you terminal will give an output similar to this:

```
Spark is listening on port 9999 and ready...
-------------------------------------------
Time: 1675241775000 ms
-------------------------------------------
(01-09-202,2)
(01-09-2022,22)
(1007,6)
(1008,5)
(50,5)
(DCardio,19)
(Lorena,5)
(Marina,5)
(Ester,6)
(DGineco,5)
...
-------------------------------------------
Time: 1675241780000 ms
-------------------------------------------
(01-09-2022,9)
(1007,3)
(1008,2)
(50,1)
(DCardio,8)
(Lorena,1)
(Marina,2)
(Ester,3)
(DGineco,1)
(Pedro,3)
...
```

Now we are going to introduce a more significant change in our program. We want to analyze only specific fields of the incoming data stream. In particular we would like to supervise in nearly real time the number of appointments by department.

Remember we are processing data in CSV format, and as you probably know, CSV files commonly incorporate a header row with the names of the columns. If we keep the code as it was in our previous examples, that header row will be inappropriately processed. Therefore, we must introduce a filter to screen out this row. As we do not know the moment in time in which this header is going to arrive to our streaming process, we have to find a way to prevent this row from being processed. As we know the header is the only row beginning with a word (alphabetical string), instead of a number, we can use a regular expression to filter rows beginning with a word.

Here is our code snippet tuned to filter the header row and to extract and process only one of the fields of interest, Department Number (DNom):

```scala
import org.apache.spark.SparkConf
import org.apache.spark.sql.SparkSession
import org.apache.spark.streaming.{Seconds, StreamingContext}
import java.io.IOException

val host = "localhost"
val port = 9999

try{
    val spark = SparkSession
    .builder()
    .master("local[*]")
    .appName("Hands-On_Spark3_socketTextStream")
    .getOrCreate()

    val sc = spark.sparkContext

    // Create the context with a 5 seconds batch size
    val ssc = new StreamingContext(sc, Seconds(5))

    val lines = ssc.socketTextStream(host, port)

    printf("\n Spark is listening on port 9999 and ready...\n")

    val filterHeaders = lines.filter(!_.matches("[^0-9]+"))
    val selectedRecords = filterHeaders.map{ row =>
    val rowArray = row.split(",")
      (rowArray(3))
      }
```

```
        selectedRecords.map(x => (x, 1)).reduceByKey(_+_).print()
        ssc.start()
        ssc.awaitTermination()
}catch {
        case e: java.net.ConnectException => println("Error establishing
        connection to " + host + ":" + port)
        case e: IOException => println("IOException occurred")
        case t: Throwable => println("Error receiving data", t)
        } finally {
        println("Finally block")
        }
```

After applying these changes, if you execute the program again and paste the following lines to your terminal

```
NSS,Nom,DID,DNom,Fecha
1004,Tomás,30,DEndo,01-09-2022
1005,Lorena,50,DGineco,01-09-2022
1006,Pedro,10,DCardio,01-09-2022
1007,Ester,10,DCardio,01-09-2022
1008,Marina,10,DCardio,01-09-2022

NSS,Nom,DID,DNom,Fecha
1009,Julia,20,DNeuro,01-09-2022
1010,Javier,30,DEndo,01-09-2022
1011,Laura,50,DGineco,01-09-2022
1012,Nuria,10,DCardio,01-09-2022
1013,Helena,10,DCardio,01-09-2022
1014,Nati,10,DCardio,01-09-2022
```

you will see an output like this:

```
-------------------------------------------
Time: 1675284925000 ms
-------------------------------------------
(DCardio,3)
(DGineco,1)
(DEndo,1)
```

```
-------------------------------------------
Time: 1675284930000 ms
-------------------------------------------
(DCardio,3)
(DGineco,1)
(DEndo,1)
(DNeuro,1)
```

As you can appreciate, header lines are removed; therefore, only the lines of interest are considered.

Next, we are going to see the other basic source directly available in the Spark Streaming core API, file systems compatible with HDFS (Hadoop Distributed File System).

File System Streaming Sources

Spark Streaming can use file systems as input data sources. File streams are used for streaming data from a folder. Spark can mount file streaming processes on any HDFS-compatible file system such as HDFS itself, AWS S3, NFS, etc. When a file system stream is set up, Spark monitors the path indicated and processes any files created in it. By default, files are processed according to the file modification timestamp, with the oldest modified files first; however, the order can be reversed using the latestFirst option (default: false), which instructs Spark to start with the latest files modified first. Spark by default supports different file formats such as text, CSV, JSON, ORC, and Parquet.

A DStream from files can be created using streamingContext.fileStream[KeyClass, ValueClass,InputFormatClass](dataDirectory), though for text files StreamingContext.textFileStream(dataDirectory) can be used. The variable dataDirectory represents the path to the folder to be monitored.

How Spark Monitors File Systems

Spark monitors file systems according to the following patterns:

- For paths such as "hdfs://hadoop:9000/folder/", "s3//…", "file//…", etc., Spark processes the files as soon as they appear under the path.

- Glob patterns to specify directories "hdfs://hadoop:9000/folder/textfiles/*/*" are also possible.

- All files within the path have to be in the same format.

- The number of files present under the path influences the time Spark will take to scan it, even if no file has been updated.

- File updates within the same time window are ignored. Therefore, once a file is processed, updating it will not cause its reprocessing.

- Spark will process files looking at the modification time, not the creation time. Therefore, files already present in the path when the streaming process starts will not be processed.

- Setting access time of a file using Hadoop FileSystem.setTimes() can cause a file to be processed out of the current processing time window.

Now we are going to see how to use Spark to continue monitoring and streaming files from a folder. In this case we are going to continue with the basic near-real-time streaming Hospital Queue Management System, tweaking it a little bit again to use it to stream files from a file system.

As we previously did with the socket data source example, two versions of the program are provided. The first one could be compiled with sbt or another Scala compiler.

The first code (*textFileStream.scala*) is shown next, and it is the Scala variant that can be compiled and executed in Spark using the *$SPARK_HOME/bin/spark-submit* command. In this case considering that we are going to pour CSV files into a folder, we should start from the last version of our previous example in which we were filtering lines beginning with a word and screening them out of the stream process:

```scala
package org.apress.handsOnSpark3

import org.apache.spark.sql.SparkSession
import org.apache.spark.streaming.{Seconds, StreamingContext}
import java.io.IOException

object textFileStream {
  def main(args: Array[String]): Unit = {

    val folder="/tmp/patient_streaming"

    try {
    val spark: SparkSession = SparkSession.builder()
```

```
.master("local[1]")
.appName("Hand-On-Spark3_textFileStream")
.getOrCreate()

spark.sparkContext.setLogLevel("ERROR")

val sc = spark.sparkContext
val ssc = new StreamingContext(sc, Seconds(5))

val lines = ssc.textFileStream(folder)

printf(f"\n Spark is monitoring the folder $folder%s and
ready... \n")

val filterHeaders = lines.filter(!_.matches("[^0-9]+"))
val selectedRecords = filterHeaders.map { row =>
val rowArray = row.split(",")
(rowArray(3))
}
selectedRecords.map(x => (x, 1)).reduceByKey(_ + _).print()
ssc.start()
ssc.awaitTermination()

} catch {
case e: IOException => println("IOException occurred")
case t: Throwable => println("Error receiving data", t)
} finally {
println("Finally block")
}
}
}
```

Pay attention to the local[1] option. In this case we have used only "[1]" because file streams do not require executing a receiver; therefore, no additional cores are required for file intake.

The next piece of code is a version of the preceding Hospital Queue Management System application that can be executed in Spark using a notebook application such as Jupyter, Apache Zeppelin, etc.:

```scala
import org.apache.spark.SparkConf
import org.apache.spark.sql.SparkSession
import org.apache.spark.streaming.{Seconds, StreamingContext}
import java.io.IOException

val folder="/tmp/patient_streaming"

try{
    val spark = SparkSession
    .builder()
    .master("local[1]")
    .appName("Hand-On-Spark3_textFileStream")
    .getOrCreate()

    val sc = spark.sparkContext

    // Create the context with a 5 seconds batch size
    val ssc = new StreamingContext(sc, Seconds(5))

    val lines = ssc.textFileStream(folder)

    printf(f"\n Spark is monitoring the folder $folder%s and
    ready... \n")

    val filterHeaders = lines.filter(!_.matches("[^0-9]+"))
    val selectedRecords = filterHeaders.map{ row =>
    val rowArray = row.split(",")
    (rowArray(3))
    }
    selectedRecords.map(x => (x, 1)).reduceByKey(_+_).print()
    ssc.start()
    ssc.awaitTermination()
}catch {
    case e: IOException => println("IOException occurred")
    case t: Throwable => println("Error receiving data", t)
    } finally {
    println("Finally block")
    }
```

Now it is time to run these examples and see their outcomes.

Running File System Streaming Applications Locally

In this case our program is going to monitor the selected /tmp/patient_streaming path and process the files copied over there as soon as Spark discovers them.

As in the "Running Socket Streaming Applications Locally" section, running the file system data source examples provided here also depends on the method you have chosen to execute them. Here you can also choose either to build your own JAR file from the code snippets provided before or use the notebook version. In any case, running the application consists of a two-step process:

1. Depending on how you are running the application

    1.1. If you are using a JAR file, open a terminal in your computer and execute your application as shown in the following:

    ```
    $SPARK_HOME/bin/spark-submit --class org.apress.handsOnSpark3.
    textFileStream --master "local[1]" /PATH/TO/YOUR/
    HandsOnSpark3-textFileStream.jar
    ```

    1.2. If you are using a notebook, just execute the code in your notebook.

2. Open a new terminal in your computer to copy the CSV files provided to the monitored folder.

    As soon as you see on your screen the message Spark is monitoring the folder /tmp/patient_streaming and ready..., you can go back to step 2 and start copying the CSV files to the /tmp/patient_streaming folder,[6] for example:

    ```
    cp /PATH/TO/patient1.csv /tmp/patient_streaming
    cp /PATH/TO/patient2.csv /tmp/patient_streaming
    cp /PATH/TO/patient3.csv /tmp/patient_streaming
    cp /PATH/TO/patient4.csv /tmp/patient_streaming
    cp /PATH/TO/patient5.csv /tmp/patient_streaming
    ```

    With a cadence of seconds, you will start seeing on your terminal session or notebook an output similar to the next one:

    ```
    Spark is monitoring the folder /tmp/patient_streaming and ready...
    ```

---

[6] It is advised to copy the files progressively to better see how Spark processes them.

```
-----------------------------------------------
Time: 1675447065000 ms
-----------------------------------------------
(DCardio,1)

-----------------------------------------------
Time: 1675447070000 ms
-----------------------------------------------
(DEndo,1)
(DNeuro,1)

-----------------------------------------------
Time: 1675447075000 ms
-----------------------------------------------
(DGastro,1)
(DCardio,3)
(DGineco,1)
(DNeuro,2)
```

3. Application termination

    Once again, `awaitTermination()` waits for a user's termination
    signal. Thus, going to the terminal session started in step 2 and
    pressing Ctrl+C or SIGTERM, the streaming context will be
    stopped. If the application is run in a notebook, you can stop
    the execution of the application by stopping or restarting the
    Spark kernel.

Known Issues While Dealing with Object Stores Data Sources

File systems such as Hadoop Distributed File System (HDFS) can establish the
modification time of its files at the beginning output stream that creates them. That is
to say, modification time can be set before the file writing process that creates them
is completed. Consequently, this behavior can cause Spark DStream to include those
incomplete files in the current processing window and subsequently ignore ulterior
data aggregations or file updates; therefore, some data can be lost as it is left out of the
window stream.

Several techniques can be used to work around this possible problem depending on the file system technology considered. In some cases direct data writing to the Spark supervised path can be fine; in others files can be created in a different folder than the final destination and finally copied or renamed to the monitored path once they are complete, though in this case file copy or rename operations also take time and the file metadata can also be altered, for example, the original file creation or modification time can be overwritten with the rename or copy time. In any case, when using Spark Streaming, attention should be paid to these details.

Advanced Input Sources

Advanced sources are not part of the StreamingContext API and are only available via third-party extra classes (similar drivers for peripherals in an operating system). Examples of these advanced sources are Kafka and Kinesis.

At the time this book was written, Spark Streaming 3.3.1 was compatible with Kafka broker versions 0.10 and higher and with Kinesis Client Library 1.2.1.

6.8 Spark Streaming Graceful Shutdown

In our previous streaming examples, we interrupted the execution of the stream process by pressing Ctrl+C or SIGTERM, thus killing the execution while it was still listening to a socket port or monitoring a file system directory. In a real production environment, a Spark streaming application cannot be abruptly interrupted because data is going to be lost in all likelihood. Imagine that while you were running our previous examples, some data stream was being received or some file was being read from the disk when you interrupted the job. Taking into consideration there is a time interval between each data ingestion and processing, if you interrupt the application in between them, that information will be lost.

In a production environment, the situation is completely different. Ideally, a Spark streaming application must be up and running 24/7. Therefore, a production job should never stop; it should constantly be reading events/files from a data source, processing them, and writing the output into a sink or another component of the data pipe.

However, in real life things are far away from being perfect. Sometimes we need to stop the streaming job for several reasons; one of them could be when a new version of our Spark application is deployed into production. In these cases, when our goal is to perform a smooth shutdown of the Spark streaming job, we have to find a way to accomplish it without data loss. It turns out there is a procedure called Graceful Shutdown that guarantees no job is forcefully halted while ongoing RDDs are processed. This part of the chapter focuses on stopping processing—gracefully.

Graceful Shutdown is a feature available in Spark Streaming to facilitate a "safe" stopping of a stream job without data loss under certain conditions we explain next. Graceful Shutdown permits the conclusion of the jobs already in progress as well as the ones piled up before closing the stream listening/reading process, and only after that the streaming job is stopped; therefore, there is no data loss under certain conditions we are going to explain later on.

To understand the Spark Streaming Graceful Shutdown feature, it is necessary to understand how Spark Streaming jobs are stopped in advance.

The logic behind Spark start and stop streaming jobs is handled by the JobScheduler as you can see in Figure 6-5.

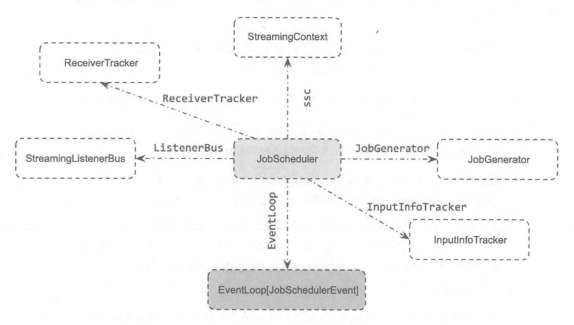

Figure 6-5. *JobScheduler and dependent services*

The Streaming scheduler or JobScheduler is created and starts with creating the StreamingContext procedure. The JobScheduler role is to track jobs submitted for execution in the jobSets internal map, which is a collection of streaming jobs.

Enabling Graceful Shutdown requires the creation of a Spark context with the parameter `spark.streaming.stopGracefullyOnShutdown` set to `true`. When Graceful Shutdown is not enabled, the JobScheduler forcefully stops stream jobs as follows:

- New data intake is prevented.

- When dynamic allocation is active, executor allocators are stopped.

- The generation of new jobs is interrupted.

- Currently executing jobs are stopped.

- Finally, job event listeners are halted.

You can see the procedure of terminating a stream process is quite "violent," representing data in transit is lost.

However, when Graceful Shutdown is enabled, the Spark JobScheduler behaves in a less radical way:

- New data intake is prevented. Graceful Shutdown waits until the receivers have received all the data in transit.

- Executor allocators are stopped. This shutdown step is not changed.

- The generation of new jobs is interrupted. The generation of new jobs is permitted, but only for the time interval in progress.

- Currently executing jobs are stopped. Graceful Shutdown sets the input parameter `processAllReceivedData` to `true`. This action grants 1 additional hour to the jobExecutor Thread Pool before termination. This parameter is not configurable, and it assumes it is time enough to finish the ongoing jobs. Otherwise, the jobExecutor Thread Pool is terminated in 2 s.

- Job event listeners are halted. This step of the stopping process is not changed either.

Now it is time to tweak our previous examples, to show how Graceful Shutdown could be implemented:

```
import org.apache.spark.SparkConf
import org.apache.spark.storage.StorageLevel
import org.apache.spark.streaming.{Seconds, StreamingContext}
```

```scala
import org.apache.spark.sql.SparkSession
import org.apache.hadoop.conf.Configuration
import org.apache.hadoop.fs.{FileSystem, Path}

val spark = SparkSession
      .builder()
      .master("local[3]")
      .appName("streamingGracefulShutdown")
      .config("spark.streaming.stopGracefullyOnShutdown", true)
      .getOrCreate()

import spark.implicits._

val sc = spark.sparkContext
val ssc = new StreamingContext(sc, Seconds(5))
val host = "localhost"
val port = 9999

val altFolder = "/tmp/alt_folder"
var stopFlag:Boolean = false

val groupedRecords =lines.map(record =>{
                              val arrayRecords=record.split(",")
                                  (arrayRecords(3))
                       })
groupedRecords.countByValue().print()

val words = lines.flatMap(_.split(","))
val wordCounts = words.map(x => (x, 1)).reduceByKey(_+_)
wordCounts.print()

ssc.start()

val timeout = 10000
var wasStopped = false

while (! wasStopped) {
      printf("\n Listening and ready... \n")
      wasStopped = ssc.awaitTerminationOrTimeout(timeout)
```

```
    if (wasStopped)
    println("Streaming process is no longer active...")
    else
    println("Streaming is in progress...")

    // Check the existence of altFolder, /tmp/alt_folder
    if (!stopFlag) {
    val fs = FileSystem.get(new Configuration())
    stopFlag = fs.exists(new Path(altFolder))
    }

    if (!wasStopped && stopFlag) {
    println("Stopping ssc context...")
    ssc.stop(stopSparkContext = true, stopGracefully = true)
    println("ssc context has been stopped!")
    }
}
```

Now, before you execute the preceding code example, in a terminal set up the socket server by typing

```
nc -lk 9999
```

After that, you can execute the preceding code snippet, and as soon as you see the message `Listening and ready...` on your screen, you can start copying and pasting the CSV example lines provided, for example:

```
1004,Tomás,30,DEndo,01-09-2022
1005,Lorena,50,DGineco,01-09-2022
1006,Pedro,10,DCardio,01-09-2022
1007,Ester,10,DCardio,01-09-2022
...
1010,Javier,30,DEndo,01-09-2022
1011,Laura,50,DGineco,01-09-2022
1012,Nuria,10,DCardio,01-09-2022
1013,Helena,10,DCardio,01-09-2022
1014,Nati,10,DCardio,01-09-2022
1009,Julia,20,DNeuro,01-09-2022
1010,Javier,30,DEndo,01-09-2022
```

In a few seconds you would see an output similar to this coming out of your program:

```
Listening and ready...
-------------------------------------------
Time: 1675631795000 ms
-------------------------------------------
(DCardio,12)
(DGineco,4)
(DEndo,4)
(DNeuro,2)

-------------------------------------------
Time: 1675631795000 ms
-------------------------------------------
(01-09-2022,22)
(1007,2)
(1008,2)
(Laura,2)
(Julia,2)
(50,4)
(Nuria,2)
(1009,2)
(DCardio,12)
(Javier,2)
...

-------------------------------------------
Time: 1675631800000 ms
-------------------------------------------

-------------------------------------------
Time: 1675631800000 ms
-------------------------------------------

Streaming in progress. Timeout...
```

```
 Listening and ready...
--------------------------------------------
Time: 1675631805000 ms
--------------------------------------------
```

Next, in a new terminal type, the following command line

```
mkdir /tmp/alt_folder
```

to create the /tmp/alt_folder folder.

Once again, in a few seconds, after the timeout period defined, you should see the following lines:

```
Stopping ssc context...
WARN ReceiverSupervisorImpl: Receiver has been stopped

--------------------------------------------
Time: 1675631815000 ms
--------------------------------------------

--------------------------------------------
Time: 1675631820000 ms
--------------------------------------------

ssc context has been stopped!

 Listening and ready...
Streaming process is no longer active...
```

Graceful Shutdown finished the job in queue and nicely stopped your streaming process without losing any data.

If you look carefully through the preceding code snippet, you can see that Spark is uninterruptedly listening to the network socket localhost:9999 (*host:port*) while the flag stopFlag is false. Thus, we need to find a way to send to Spark the stop streaming signal. We achieve that by creating a new folder in the defined file system path /tmp/alt_folder.

The next two lines of code

```
val fs = FileSystem.get(new Configuration())
stopFlag = fs.exists(new Path(altFolder))
```

permit checking whether the path defined by the altFolder variable exists. If it exists, the stopFlag Boolean variable is set to true and hence triggers the Grateful Shutdown process.

6.9 Transformations on DStreams

Similar to those on RDDs, transformations on DStreams allow the data from the input DStream to be transformed according to our needs.

Apart from the transformations on DStreams already seen in previous examples, DStreams support several other transformations available on normal Spark RDDs. Some of the common transformations on DStreams are listed and succinctly explained in Table 6-1.

Table 6-1. *Transformations on DStreams*

Transformation	Description
map(func)	Returns a new DStream after applying a function func to all the DStream elements.
flatMap(func)	Similar to map, it takes a function as an argument and applies it to each element, returning a new RDD with 0 or multiple elements.
filter(func)	Returns a new DStream screening out the incoming DStream records on which the function func returned false.
repartition(numPartitions)	Changes the DStream level of parallelism, increasing or decreasing the number of partitions.
union(otherStream)	Returns a DStream union of the elements of the source DStream and otherDStream.
count()	Returns a new DStream counting the number of elements of a DStream.
reduce(func)	Returns a new DStream by aggregating the elements in each RDD of a DStream using a function func. The function should support parallelized computation.
countByValue()	Returns a new DStream of (K, Long) pairs with the frequency in each key in the RDD of the DStream.
reduceByKey(func, [numTasks])	Returns a DStream of (Key, Value) pairs aggregating the values for each key using a reduce function.
join(otherStream, [numTasks])	Returns a new DStream of (K, (V, W)) pairs from a joining operation of two DStreams of (K, V) and (K, W) key-value pairs.
cogroup(otherStream, [numTasks])	Returns a DStream of (K, Seq[V], Seq[W]) tuples from a DStream of (K, V) and (K, W) pairs.
transform(func)	Returns a DStream after applying arbitrary RDD-to-RDD functions to a DStream.
updateStateByKey(func)	Returns a "state" DStream maintaining it updated with new information from previous DStreams.

6.10 Summary

In this chapter we have explained what Apache Spark Streaming is, together with the Spark DStream (Discretized Stream) as the basic abstraction behind the Spark Streaming concept. We mentioned DStream is a high-level abstraction for Spark Streaming just like RDD. We also went through the differences between real-time analytics of bound and unbound data, mentioning the challenges and uncertainties stream processing brings in. Next, we talked about the Spark Streaming Execution Model and stream processing architectures. At that point, we explained the Lambda and Kappa architectures as the main stream processing architectures available. After that, we went through the concepts of Discretized Streams and stream sources and receivers. The last point was quite dense, explaining and giving examples of basic and advanced data sources. The advanced topic of Grateful Shutdown was described, giving a practical example, and finally, a list of the most common transformations on DStreams was provided.

PART II

Apache Spark Streaming

CHAPTER 7

Spark Structured Streaming

Nowadays, in the big data world, more and more business processes and daily used applications require the analysis of real-time or near-real-time information at scale. Real-time data analysis is commonly associated with processes that require decisions to be taken quickly and without delay. Therefore, infrastructures capable of providing instant analytics, management of continuously flowing data, and fault tolerance and handling stragglers or slow components are necessary.

Considering the main characteristics that define data streaming, in which

- Information is continuous

- Information is unbounded

- There is high volume and velocity of data production

- Information is time-sensitive

- There is heterogeneity of data sources

we can assume data faults and stragglers[1] are certain to occur in this sort of environment.

Data faults and stragglers represent a serious challenge for streaming data processing. For instance, how can we get insights from a sequence of events arriving at a stream processing system if we do not know what is the order in which they took place?

---

[1] Late or out-of-order events (information).

© Alfonso Antolínez García 2023
A. Antolínez García, *Hands-on Guide to Apache Spark 3*, https://doi.org/10.1007/978-1-4842-9380-5_7

Stream processing uses timestamps[2] to sequence the events and includes different notions of time regarding stream event processing:

- *Event time*: It corresponds with the moment in time in which the event is generated by a device.

- *Ingestion time*: It is the time when an event arrives at the stream processing architecture.

- *Processing time*: It refers to the computer time when it begins treating the event.

In Chapter 6 we saw that Spark's first attempt to keep up with the dynamic nature of information streaming and to deal with the challenges mentioned before was the introduction of Apache Spark Streaming (DStream API). We also studied that DStreams or Discretized Streams are implemented on top of Spark's Resilient Distributed Dataset (RDD) data structures. DStream handles continuous data flowing by dividing the information into small chunks, processing them later on as micro-batches.

The use of the low-level RDD API offers both advantages and disadvantages. The first disadvantage is that, as the name states, it is a low-level framework; therefore, it requires higher technical skills to take advantage of it and poses performance problems because of data serialization and memory management. Serialization is critical for distributed system performance to minimize data shuffling across the network; therefore, if not managed with caution, it can lead to numerous issues such as memory overuse and network bottlenecks.

7.1 General Rules for Message Delivery Reliability

At this point we provide an introduction to general message delivery semantics you are going to find in the next chapters and the importance of each one in a streaming data processing infrastructure.

Performing complex real-time streaming analytics is not an easy task. In order to give some context to the importance of message delivery reliability in real-time streaming analytics, consider a streaming application collecting real-time events from remote hosts or sensors, which can generally be called actors, scattered over multiple

---

[2] The point in time at which an event takes place.

locations. Also consider these actors are going to be connected to different network topologies and to different qualities of hardware with observable differences in message latency (the time it takes for messages to travel from one point on a network to another), bandwidth (the capacity for data transfer of an electronic communications system), and reliability. Therefore, the more steps are involved in the event transmission, the more likely the sequence of messages can be faulty.

When all of these factors are taken into account, we conclude that regarding real-time data streaming processing, we can only rely on those properties that are always guaranteed in order to achieve full actor's location transparency and strict warranties on message delivery.

When it comes down to the semantics of message delivery reliability mechanisms, there are three following basic categories:

- *At most once (at-most-once delivery)*: This semantic means the message is delivered once or not at all (in a fire-and-forget manner). This means that the message can be lost. It is the most inexpensive in terms of highest delivery performance and least implementation overhead as no state is kept either during the sending process or during the transport process. As it gets rid of the overhead of waiting for acknowledgment from the message brokers, it is suitable for use cases in which attaining a high throughput is of paramount importance and when losing some messages does not significantly affect the final result, for example, analyzing customer sentiment by listening to posts in social networks. Among millions of them, losing a few will not probably greatly impact the final conclusions.

- *At least once (at-least-once delivery)*: In this semantic, multiple attempts are going to possibly be made in order to guarantee that at least one message reaches the destination. It is suitable for use cases in which there is little or no concern with duplication of data but it is of utmost importance no message is lost, for example, sensors monitoring critical components or human vital signs (body temperature, pulse rate, respiration rate, blood pressure, etc.).

- *Exactly once (exactly-once delivery)*: This semantic means messages can neither be lost nor duplicated. Messages are delivered and read once. Therefore, exactly one delivery is made to the destination.

Even though there is a lot of literature out there about the impossibility of achieving an exactly-once delivery as it implies the guarantee that a message is delivered to the recipient once, and only once, from a theoretical point of view, it exists, and we consider it in this book.

7.2 Structured Streaming vs. Spark Streaming

Considering only the Spark consolidated versions and modules, both Spark Streaming and Spark Structured Streaming use the default micro-batch processing model; however, while Spark Streaming employs DStreams, Spark Structured Streaming uses datasets/DataFrames. A DStream is represented by a perpetual sequence of RDDs, which are Spark's notion of immutable, distributed datasets that are held in memory. DStream relies on RDDs to provide low-level transformation and processing. On the other hand, Structured Streaming takes advantage of the DataFrame and Dataset APIs, providing a higher level of abstraction and permitting SQL-like manipulation functions. As RDDs are part of the low-level API, they can only work with event ingestion time, also known as processing time or the time when the event entered the engine, therefore being unable to efficiently tackle out-of-order events. Conversely, Structured Streaming can process data based on the event time, the time when the event was generated, hence providing a workaround to deal with received late and out-of-order events. Spark RDDs cannot be optimized; therefore, they are more likely to develop inefficient data transformations, and optimization would require extra work from the programmer side.

Additionally, using DStream is not straightforward to build in-stream processing pipelines supporting exactly-once-guarantee delivery policies. Implementation is possible, but requires programming workarounds. In contrast, Structured Streaming incorporates new valuable concepts for in-stream processing:

- The exactly-once-guarantee message delivery rule is implemented by default; therefore, theoretically, data is processed only once, and duplicates are removed from the outcomes.

- Event-time in-stream-based processing brings the benefits mentioned before.

Another important difference between using DStreams and Structured Streaming is necessity or not of a streaming sink. While DStream streaming outputs are RDDs that can be manipulated and, hence, do not have a need for a streaming sink as final output, Spark Structured Streaming requires a streaming sink.

Another one is how real-time streaming is treated. While DStream simulates real-time processing stockpiling data into micro-batches, Spark Structured Streaming uses the concept of unbounded table, which we will explain in detail later on in this chapter, to continuously add real-time events to the streaming flow.

Another significant difference between Spark DStreams and Spark Structured Streaming is how the end-to-end streaming process is conducted. On one hand, Spark Structured Streaming has a dedicated thread to check whether new data has arrived to the stream, and if and only if there is new information to process, the stream query is executed. On the other hand, while a Spark Streaming program is running, DStream's micro-batches are executed according to the `batchDuration` time interval parameter of the `StreamingContext()` method at which the DStream generates a RDD independently of there is live information or not.

7.3 What Is Apache Spark Structured Streaming?

Apache Spark Structured Streaming was introduced with Spark 2.0. Spark Structured Streaming is a scalable and near-real-time stream processing engine offering end-to-end fault tolerance with exactly-once processing guarantees. Spark Structured Streaming is built on top of the Spark SQL library; hence, it natively incorporates the Spark SQL code and memory optimization and facilitates the use of SQL. Structured Streaming is based on the Dataframe and Dataset APIs.

Spark Structured Streaming incorporates several novel conceptualizations where the most important ones are the following:

- Input table

- Result table

- Output modes

- Datasets and DataFrames Streaming API

- Event-Time Window Operations

- Watermarking

- Streaming data deduplication

- State store (a versioned key-value store)

- Output sinks

Let's explain next what is behind some of these concepts. The others will be explained in later chapters.

Spark Structured Streaming Input Table

Spark Structured Streaming uses the concept of "input table," which could be assimilated to the "unbounded input table" abstraction depicted in Figure 7-1, to process every input data. The concept of unbounded table means that every new piece of data arriving at the system is appended as a new row to the table.

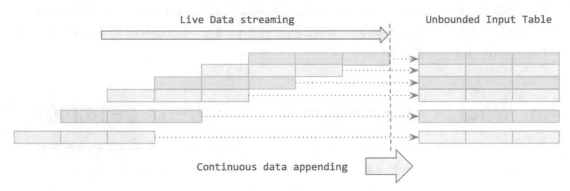

Figure 7-1. *The Spark Structured Streaming unbounded input table flow diagram*

Consequently, with Structured Streaming, computation is performed incrementally with continuous result update as data comes in, permitting the representation of stream data processing in the same fashion batch computation on at-rest data is represented.

Spark Structured Streaming Result Table

The "result table" in Spark Structured Streaming could also be assimilated to a kind of unbounded output table. The result table will eventually be the consequence of every query on the input data. Every time new information is added to the unbounded input table, it will trigger the update of the unbounded output table, consequently writing the results to the designated output or data sink according to the out mode established. The unbounded output table concept and how it integrates into the streaming workflow are depicted in Figure 7-2.

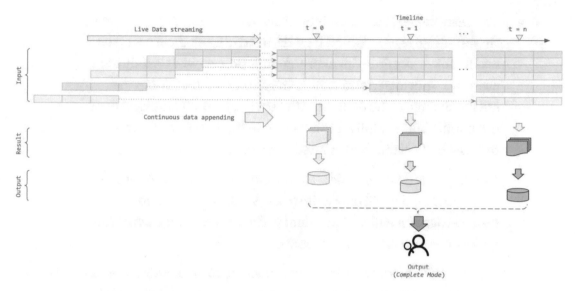

Figure 7-2. *The Spark Structured Streaming unbounded output table flow diagram*

Next, we explain the three different output modes available in Spark Structured Streaming.

Spark Structured Streaming Output Modes

The output mode (alias: OutputMode) is a new concept introduced by Structured Streaming, and as we have already mentioned before, Spark Structured Streaming requires a streaming sink.

When it comes to Spark Structured Streaming, the output mode specifies what data is written to a streaming sink and the way of writing that data.

Spark Structured Streaming supports three output modes:

- *Append mode* (alias: append). This is the default behavior, and only the *new rows* that arrived at the result table are written to the output sink. Regarding streaming aggregations, *new rows* are those whose intermediate states become final. This mode guarantees that each row will be output only once; it is indicated when we are only interested in analyzing the new data. Append mode is applicable to only queries where rows appended to the result table are not going to be modified, for example, those only employing select, where, map, flatMap, filter, join, etc.

- *Complete mode* (alias: complete). This mode is only supported for streaming aggregations, and it works by writing all the rows of the result table every time the information is processed. It is indicated when we want to perform data aggregation and then dump the full result table to the data sink after every update. This mode does not handle how the full table data is written; therefore, it is the responsibility of the used connector to manage the information.

- *Update mode* (alias: update). This output mode was introduced with Spark 2.1.1, and it will only write to the sink the updated rows since the last trigger. It will be equivalent to the append mode when the query does not contain aggregations.

However, no output mode mentioned is applicable to all possible streaming queries. Table 7-1 shows an output mode vs. streaming query compatibility matrix.[3]

Table 7-1. *Output Mode vs. Streaming Query Compatibility Matrix*

Streaming Query Type		Supported Output Modes
Queries with aggregation	Aggregation in event time with watermark	Append, Update, Complete
	Other aggregations	Complete, Update
Queries with `mapGroupsWithState`		Update
Queries with `flatMapGroupsWithState`	Append operation mode	Append
	Update operation mode	Update
Queries including `joins`		Append
Other queries		Append, Update

The output mode is specified on the writing side of a streaming query using the streaming.DataStreamWriter.outputMode method using either an alias or a value of the org.apache.spark.sql.streaming.OutputMode object:

---

[3] Based on the official Apache Spark Structured Streaming Programming Guide.

```scala
import org.apache.spark.sql.SparkSession
import org.apache.spark.sql.streaming.OutputMode.Update

val spark:SparkSession = SparkSession.builder()
  .master("local[*]")
  .appName("Hands-On-Spark3_Structured_Streaming")
  .getOrCreate()

val inputStream = spark
  .readStream
  .format("socket")
  .option("host","localhost")
  .option("port","9999")
  .load()

inputStream.select(explode(split(df("value")," "))
  .alias("palabra"))
  .groupBy("word")
  .count()
  .writeStream
  .format("console")
  .outputMode("complete") // Complete output mode selected
  .start()
  .awaitTermination()

// Another way of specifying

val inputStream = spark
  .readStream
  .format("socket")
  .load()
// .... your code goes here
  .writeStream
  .format("console")
  .outputMode(Update) // Update output mode selected
  .start()
  .awaitTermination()
```

Remember when using Structured Streaming, you always have to specify a streaming sink and output mode.

In summary, taking into consideration the Spark Structured Streaming characteristics enumerated before, we can say that Structured Streaming offers consistency in the outcomes provided as it is guaranteed they are always going to be equivalent to the ones returned by an equivalent batch process having the same input data. However, each of these modes is applicable only to certain types of queries.

7.4 Datasets and DataFrames Streaming API

Since Spark 2.0, DataFrames and datasets can be used to process both data at rest (bound data) and streaming data (unbound data). The most recent Spark versions can use DataFrames/datasets to process data coming from streaming sources using the common SparkSession entry point and apply exactly the same transformations and actions to them that could be applied in a batch process.

Streaming DataFrames can be created via the DataStreamReader interface returned by `SparkSession.readStream()` if you are using Scala, Java, or Python or the `read.stream()` method if you are using R, and like the Spark SQL read() method to read different format files into a Spark DataFrame, the details of the source data format (CSV, JSON, etc.), schema, etc. can be specified.

As with Spark Streaming (DStreams), Structured Streaming also incorporates some built-in data sources for data ingestion. These data sources are

- *File source*: Streaming data is uploaded from different format files such as text, CSV, JSON, Parquet, etc. located in a directory using the DataStreamReader method.

- *Kafka source*: Streaming data is read from Apache Kafka topics. At the time this book was written, only Kafka broker version 0.10.0 or higher was supported.

- *Socket source*: Reads UTF-8-encoded text from a socket connection. It is not considered a fault-tolerant streaming source as it does not support using checkpointed offsets to resubmit data after a failure occurs.

- *Rate source*: This streaming source is intended for testing and benchmarking only. It is used to produce random data with two columns, "timestamp" and "value," at the specified number of rows per second (rate). The "timestamp" column contains the time the message was sent out in a Timestamp-type format, while the "value" column allocates a Long-type number counting messages sent and starting from 0.

- *Rate per micro-batch source*: This streaming data source is also intended for testing and benchmarking only. As it happens with the rate source, each output row contains two columns, "timestamp" and "value." It has the same characteristics as the rate source, but unlike the latter, the former is intended to provide a consistent number of rows per micro-batch (timestamp and value), that is to say, if batch 0 produces numbers 0 to 999 and their associated timestamps, batch 1 will produce 1000 to 1999 and their subsequent timestamps, and so on.

Next, in Table 7-2 you can see a summary of the above-mentioned data sources and their main options. This table again is based on the official Apache Spark input sources for streaming DataFrames and datasets.[4]

---

[4] https://spark.apache.org/docs/latest/structured-streaming-programming-guide.html#input-sources

Table 7-2. *Spark Structured Streaming Data Source Options*

Source	Options	Fault- Tolerant

File source	• `path`: Path to the input file directory.
	• `maxFilesPerTrigger`: Maximum number of new files to be considered in every trigger (default: no max).
	• `latestFirst`: Whether to process the latest files first.
	• `fileNameOnly`: Whether to check new files based on only the filename instead of on the full path:
	  • Default: false.
	  • Set `fileNameOnly` to "true," and the following files "hdfs://<host>:<port>/file.txt", "file:///file.txt", "s3://x/file.txt", etc. will be considered as the same because their filenames are "file.txt".
	• `maxFileAge`: Maximum age of a file in a directory before it is ignored:
	  • For the first batch, all files will be considered valid.
	  • The max age is specified with respect to the timestamp of the latest file and not the timestamp of the current system (default: 1 week).
	  • If `latestFirst` is set to "true" and `maxFilesPerTrigger` is set, then this parameter will be ignored, because old files that are valid, and should be processed, may be ignored.
	• `cleanSource`: To clean up completed files after processing. Available options are "archive," "delete," and "off" (default: off):
	  • When "archive" is provided, the `sourceArchiveDir` option must be provided as well.
	  • Both archiving (moving files) and deleting completed files will introduce overhead (slowdown) in each micro-batch.
	See the input sources manual for more details.
		Yes

(continued)

Table 7-2. (*continued*)

Source	Options	Fault- Tolerant
Socket source	Host and port (`<host>:<port>`) to connect to.	No
Rate source	• `rowsPerSecond`: How many rows should be generated per second, for example, 100 (default: 1). • `rampUpTime`: How long to ramp up before the generating speed becomes `rowsPerSecond`. Maximum granularity is seconds, for example, 5 s (default: 0 s). • numPartitions: The partition number for the generated rows, for example, 10 (default: Spark's default parallelism). The number of `rowsPerSecond` is not guaranteed as the query may be resource constrained; in that case `numPartitions` can be modified to help reach the desired rate.	Yes
Rate per micro- batch source	• `rowsPerBatch`: Number of rows that should be generated per micro-batch, for example, 100 • `numPartitions`: The partition number for the generated rows, for example, 10 (default: Spark's default parallelism) • `startTimestamp`: Starting value of generated time, for example, 1000 (default: 0) • `advanceMillisPerBatch`: The amount of time being advanced in generated time on each micro-batch, for example, 1000 (default: 1000)	Yes
Kafka source	Check out the Kafka Integration Guide documentation for more details.	Yes

Now that we have studied the basics of Spark Structured Streaming and the main sources of data, it is time to see how streaming DataFrames work with some examples.

Socket Structured Streaming Sources

To show how Spark Structured Streaming can be used to ingest data by listening to a socket connection, we are going to continue using our basic near-real-time streaming Hospital Queue Management System shown in the previous chapter, tweaking it a little bit to make it more realistic implementing a JSON input format.

We show two options to see our program up and running. The first code (*readStreamSocket.scala*) is shown next, a Scala variant that can be compiled and executed in Spark using the *$SPARK_HOME/bin/spark-submit* command. It is out of the scope of this book to discuss how to compile and link Scala code, but it is recommended to use sbt together with sbt-assembly to create a so-called *"fat JAR"* file including all the necessary libraries, a.k.a. *dependencies*:

```scala
package org.apress.handsOnSpark3

import org.apache.spark.sql.SparkSession
import org.apache.spark.sql.functions._
import org.apache.spark.sql.types.{IntegerType, StringType, StructField,
StructType}
import java.io.IOException

case class Patient(
                    NSS: String,
                    Nom: String,
                    DID: Option[Long],
                    DNom: String,
                    Date: String
                  )

object readStreamSocket {
  def main(args: Array[String]): Unit = {

    val PatientsSchema = StructType(Array(
    StructField("NSS", StringType),
    StructField("Nom", StringType),
    StructField("DID", IntegerType),
    StructField("DNom", StringType),
    StructField("Date", StringType))
    )
    val host = "localhost"
    val port = 9999

    try {
    val spark: SparkSession = SparkSession.builder()
    .master("local[*]")
```

```scala
    .appName("Hand-On-Spark3_Socket_Data_Source")
    .getOrCreate()

spark.sparkContext.setLogLevel("ERROR")

// Set up spark.readStream …
import spark.implicits._

val PatientDS = spark.readStream
.format("socket")
.option("host", host)
.option("port", port)
.load()
.select(from_json(col("value"),PatientsSchema).as("patient"))
.selectExpr("patient.*")
.as[Patient]

printf("\n Listening and ready... \n")

val selectDF = PatientDS.select("*")

selectDF.writeStream
.format("console")
.outputMode("append")
.option("truncate", false)
.option("newRows", 30)
.start()
.awaitTermination()

} catch {
case e: java.net.ConnectException => println("Error establishing
connection to " + host + ":" + port)
case e: IOException => println("IOException occurred")
case t: Throwable => println("Error receiving data", t)
} finally {
println("Finally block")
}
  }
}
```

The next piece of code is a version of the preceding Hospital Queue Management System application that can be executed in Spark using a notebook application such as Jupyter, Apache Zeppelin, etc., which can be more convenient for learning purposes, especially if you are not familiar with Scala code compiler tools:

```scala
import org.apache.spark.sql.SparkSession
import org.apache.spark.sql.functions._
import org.apache.spark.sql.types.{IntegerType, StringType, StructField, StructType,DoubleType,LongType}
import org.apache.spark.sql.{DataFrame, Dataset, Encoders, SparkSession}
import java.io.IOException

val PatientsSchema = StructType(Array(
     StructField("NSS", StringType),
     StructField("Nom", StringType),
     StructField("DID", IntegerType),
     StructField("DNom", StringType),
     StructField("Fecha", StringType))
     )

case class Patient(
     NSS: String,
     Nom: String,
     DID: Option[Long],
     DNom: String,
     Fecha: String
)

val spark:SparkSession = SparkSession.builder()
     .master("local[*]")
     .appName("Hand-On-Spark3_Socket_Data_Source")
     .getOrCreate()

spark.sparkContext.setLogLevel("ERROR")

val host = "localhost"
val port = 9999
```

```scala
try {
      val PatientDS = spark.readStream
      .format("socket")
      .option("host",host)
      .option("port",port)
      .load()
      .select(from_json(col("value"), PatientsSchema).as("patient"))
      .selectExpr("Patient.*")
      .as[Patient]

      printf("\n Listening and ready... \n")

      val selectDF = PatientDS.select("*")

      selectDF.writeStream
      .format("console")
      .outputMode("append")
      .option("truncate",false)
      .option("newRows",30)
      .start()
      .awaitTermination()
} catch {
      case e: java.net.ConnectException => println("Error establishing
      connection to " + host + ":" + port)
      case e: IOException => println("IOException occurred")
      case t: Throwable => println("Error receiving data", t)
}finally {
      println("In finally block")
}
```

Notice how we have defined the PatientsSchema schema before ingesting the data:

```scala
val PatientsSchema = StructType(Array(
      StructField("NSS", StringType),
      StructField("Nom", StringType),
      StructField("DID", IntegerType),
      StructField("DNom", StringType),
      StructField("Fecha", StringType))
   )
```

When we use Spark Structured Streaming, it is mandatory to define the schema of the information before using it.

Pay attention also to the `local[*]` option. In this case we have used "*"; thus, the program is going to use all the cores available. It is important to use more than one because the application must be able to run two tasks in parallel, listening to a TCP socket (localhost:9999) and, at the same time, processing the data and showing it on the console.

Running Socket Structured Streaming Applications Locally

We are going to use a featured networking utility called Netcat to set up a simple client/server streaming connection. Netcat (netcat, nc, ncat, etc. depending on the system) is available in Unix-like operating systems and uses the TCP/IP to read and write data through a network. In this book we use the Netcat OpenBSD version (nc).

The syntax for the nc command is

```
nc [<options>] <host> <port>
```

Netcat has several [<options>]; however, we are going to use only -l, which instructs nc to listen on a UDP or TCP <port>, and -k, which is used in listen mode to accept multiple connections. When <host> is omitted, nc listens to all the IP addresses bound to the <port> given.

To illustrate how the program works, we are going to take advantage of the nc utility introduced before, to establish a streaming client/server connection between nc and our Spark application. In our case nc will act as a server (listens to a host:port), while our application will act as a client (connects to the nc server).

Whether you have built your JAR file from the previous code or are using the notebook version, running the application consists of a two-step process:

1. Open a terminal in your system and set up the server side of the client/server streaming connection by running the following code:

    ```
    nc -lk 9999
    ```

2. Depending on how you are running the application

2.1. *Using a JAR file*: Open a second terminal and execute your application as shown in the following:

```
$SPARK_HOME/bin/spark-submit --class org.apress.
handsOnSpark3.readStreamSocket --master "local[*]"
PATH/TO/YOUR/HandsOnSpark3-readStreamSocket.jar
```

2.2. *Using a notebook*: Just execute the code in your notebook.

As soon as you see the message `Listening and ready…` on your screen, you can go back to step 1 and type some of the JSON strings provided, for example:

{"NSS":"1234","Nom":"María", "DID":10, "DNom":"Cardio", "Fecha":"01-09-2022"}
{"NSS":"2345","Nom":"Emilio", "DID":20, "DNom":"Neuro", "Fecha":"01-09-2022"}
{"NSS":"3456","Nom":"Marta", "DID":30, "DNom":"Endo", "Fecha":"01-09-2022"}
…
{"NSS":"4567","Nom":"Marcos", "DID":40, "DNom":"Gastro", "Fecha":"01-09-2022"}
{"NSS":"5678","Nom":"Sonia", "DID":50, "DNom":"Gineco", "Fecha":"01-09-2022"}
{"NSS":"6789","Nom":"Eduardo", "DID":10, "DNom":"Cardio", "Fecha":"01-09-2022"}

With a cadence of seconds, you will see an output like the following one coming up on your terminal:

```
Listening and ready...

----------------------------------------------
Batch: 1
----------------------------------------------
+----+------+---+------+----------+
|NSS |Nom   |DID|DNom  |Fecha     |
+----+------+---+------+----------+
|1234|María |10 |Cardio|01-09-2022|
|2345|Emilio|20 |Neuro |01-09-2022|
```

```
|3456|Marta |30 |Endo  |01-09-2022|
|4567|Marcos|40 |Gastro|01-09-2022|
|5678|Sonia |50 |Gineco|01-09-2022|
+----+------+---+------+----------+


-------------------------------------------
Batch: 2
-------------------------------------------
+----+-------+---+------+----------+
|NSS |Nom    |DID|DNom  |Fecha     |
+----+-------+---+------+----------+
|6789|Eduardo|10 |Cardio|01-09-2022|
+----+-------+---+------+----------+


-------------------------------------------
Batch: 3
-------------------------------------------
+----+------+---+------+----------+
|NSS |Nom    |DID|DNom  |Fecha     |
+----+------+---+------+----------+
|1009|Julia |20 |Neuro |01-09-2022|
|1010|Javier|30 |Endo  |01-09-2022|
|1011|Laura |50 |Gineco|01-09-2022|
|1012|Nuria |10 |Cardio|01-09-2022|
|1013|Helena|10 |Cardio|01-09-2022|
+----+------+---+------+----------+
```

3. Application termination

   awaitTermination() waits for a user's termination signal. Thus, going to the terminal session started in step 1 and pressing Ctrl+C or SIGTERM, the streaming context will be stopped.

This way of terminating a streaming application is neither elegant nor correct, because the operations in progress when we terminate it are going to be lost. A more elegant and correct approach is using Spark Structured Streaming Graceful Shutdown we saw in the previous chapter. Thus, we encourage you to play with the previous code adding to it the Graceful Shutdown feature.

File System Structured Streaming Sources

Spark Streaming can use file systems as input databases. Spark can mount file streaming processes on any HDFS-compatible file system such as HDFS itself, AWS S3, NFS, etc. When a file system stream is set up, Spark monitors the path indicated and processes any files created in it.

Spark monitors file systems according to the following patterns:

- For paths such as "hdfs://hadoop:9000/folder/", "s3//...", "file//...", etc., Spark processes the files as soon as they appear under the path.

- Glob patterns to specify directories "hdfs://<hadoop-host>:<hadoop-port>/folder/textfiles/*/*" are also possible.

- All files within the path have to be in the same format.

- The number of files present under the path influences the time Spark will take to scan it, even if no file has been updated.

- File updates within the same time window are ignored. Therefore, once a file is processed, updating it will not cause its reprocessing.

- Spark will process files looking at the modification time, not the creation time. Therefore, files already existing in the path when the streaming process starts will not be processed.

- Setting access time of a file using Hadoop FileSystem.setTimes() can cause a file to be processed out of the current processing time window.

Now we are going to see how to use Spark to continue monitoring and streaming files from a folder. In this case we are going to continue with the basic near-real-time streaming Hospital Queue Management System, tweaking it a little bit again to use it to stream files from a file system.

As we previously did with the socket data source example, two versions of the program are provided. The first one could be compiled with sbt or another Scala compiler:

```scala
package org.apress.handsOnSpark3

import org.apache.spark.sql.SparkSession
import org.apache.spark.sql.functions._
import org.apache.spark.sql.types.{IntegerType, StringType, StructField, StructType}
import java.io.IOException

object dStreamsFiles {
  def main(args: Array[String]): Unit = {

    val PatientsSchema = StructType(Array(
    StructField("NSS", StringType),
    StructField("Nom", StringType),
    StructField("DID", IntegerType),
    StructField("DNom", StringType),
    StructField("Date", StringType)
    )
    )

    try {
    val spark: SparkSession = SparkSession
    .builder()
    .master("local[3]")
    .appName("Hand-On-Spark3_File_Data_Source")
    .getOrCreate()

    spark.sparkContext.setLogLevel("ERROR")

    val df = spark.readStream
    .schema(PatientsSchema).json("/tmp/patient_streaming")

    val groupDF = df.select("DID")
    .groupBy("DID").agg(count("DID").as("Accumulated"))
    .sort(desc("Accumulated"))
```

```scala
        printf("\n Listening and ready... \n")

        groupDF.writeStream
        .format("console")
        .outputMode("complete")
        .option("truncate", false)
        .option("newRows", 30)
        .start()
        .awaitTermination()

        } catch {
            case e: IOException => println("IOException occurred")
            case t: Throwable => println("Error receiving data", t)
        } finally {
        println("Finally block")
        }
    }
}
```

Next is the second version intended to be executed in a notebook such as Apache Zeppelin, Jupyter Notebook, etc.:

```scala
import org.apache.spark.sql.SparkSession
import org.apache.spark.sql.types.{IntegerType, StringType, StructField,
StructType}
import org.apache.spark.sql.functions.desc
import java.io.IOException

val spark:SparkSession = SparkSession
        .builder()
        .master("local[3]")
        .appName("Hand-On-Spark3_File_Data_Source")
        .getOrCreate()

spark.sparkContext.setLogLevel("ERROR")

val PatientsSchema = StructType(Array(
        StructField("NSS", StringType),
        StructField("Nom", StringType),
```

```scala
    StructField("DID", IntegerType),
    StructField("DNom", StringType),
    StructField("Fecha", StringType)
    )
    )

try{
    val df = spark.readStream.schema(PatientsSchema)
            .json("/tmp/patient_streaming")

    val groupDF = df.select("DID")
    .groupBy("DID").agg(count("DID").as("Accumulated"))
    .sort(desc("Accumulated"))

    groupDF.writeStream
    .format("console")
    .outputMode("complete")
    .option("truncate",false)
    .option("newRows",30)
    .start()
    .awaitTermination()

} catch{
    case e: IOException => println("IOException occurred")
    case t: Throwable => println("Error receiving data", t)
}
```

Running File System Streaming Applications Locally

In this case our program is going to monitor the selected /tmp/patient_streaming path and process the files copied there as soon as Spark discovers them.

As in the "Running Socket Structured Streaming Applications Locally" section, running the file system data source examples provided here also depends on the method you have chosen to execute them. Here you can also choose either to build your own JAR file from the code snippets provided before or use the notebook version. In any case, running the application consists of a two-step process:

1. Depending on how you are running the application

    1.1. If you are using a JAR file, open a terminal in your computer and execute your application as shown in the following:

    ```
    $SPARK_HOME/bin/spark-submit --class org.apress.
    handsOnSpark3.dStreamsFiles --master "local[*]"
    target/scala-2.12/HandsOnSpark3-dStreamsFiles-assembly-
    fatjar-1.0.jar
    ```

    1.2. If you are using a notebook, just execute the code in your notebook.

2. Open a new terminal in your computer to copy the JSON files provided to the monitored folder.

    As soon as you see on your screen the message Listening and ready..., you can go back to step 2 and start copying JSON files to the /tmp/patient_streaming folder,[5] for example:

    ```
    cp /PATH/TO/patient1.json /tmp/patient_streaming
    cp /PATH/TO/patient2.json /tmp/patient_streaming
    cp /PATH/TO/patient3.json /tmp/patient_streaming
    cp /PATH/TO/patient4.json /tmp/patient_streaming
    cp /PATH/TO/patient5.json /tmp/patient_streaming
    ```

With a cadence of seconds, you will start seeing on your terminal session or notebook an output like this:

```
Listening and ready...
---------------------------------------------
Batch: 0
---------------------------------------------
+---+-----------+
|DID|Accumulated|
+---+-----------+
|10 |1          |
+---+-----------+
```

---

[5] It is advised to copy the files progressively to better see how Spark processes them.

269

```
-------------------------------------------
Batch: 1
-------------------------------------------
+---+-----------+
|DID|Accumulated|
+---+-----------+
|20 |1          |
|10 |1          |
|30 |1          |
+---+-----------+

-------------------------------------------
Batch: 2
-------------------------------------------
+---+-----------+
|DID|Accumulated|
+---+-----------+
|10 |4          |
|20 |3          |
|40 |1          |
|50 |1          |
|30 |1          |
+---+-----------+

-------------------------------------------
Batch: 3
-------------------------------------------
+---+-----------+
|DID|Accumulated|
+---+-----------+
|10 |7          |
|20 |3          |
|50 |2          |
|30 |2          |
|40 |1          |
+---+-----------+
```

The examples provided generate untyped DataFrames; it means that the schema provided is not validated at compile time, only at runtime when the code is executed.

So far, we have only been applying transformations to the data arriving to the streaming process. For example, in our last program, the "Accumulated" column adds up the input "DID" field. This is what is called stateless streaming. Suppose now a scenario in which you want to find out the total occurrences of each value received by your streaming application, updating the state of the previously processed information. Here is where the concepts of streaming state and stateful streaming that we are going to see next come into play.

7.5 Spark Structured Streaming Transformations

In this section we are going to walk you through the Spark Structured Streaming supported data transformations. These data transformations are classified as *stateless* and *stateful* operations. Only operations that can incrementally update DataFrame aggregations are supported by Spark Structured Streaming. Stateful operations need to maintain the event state across the streaming process or as long as we define.

Next, we explain the Structured Streaming notions mentioned before of streaming state and stateless and stateful operations.

Streaming State in Spark Structured Streaming

The state is one of the most important components of any streaming data pipeline. Depending on the specific use case, it might be necessary to maintain the state of different variables like counters while the streaming application is in operation.

In the process of dealing with stream data processing, every application must use either stateless or stateful data management approaches. The main difference between stateless and stateful operations is whether when executing incremental aggregations/operations we need to keep track of ongoing event states.

Summarizing, in this context *state* basically means "ephemeral information" that needs to be retained for a certain period of time in order to use it down the stream process.

Spark Stateless Streaming

Stateless state means that data included in the ongoing micro-batches is processed without considering past or future information. One example of this kind of stateless data processing could be the recollection and saving of web events (clicks, page visits, etc.).

Stateless operations are those that can process the information independently of the previously processed information. Examples of these operations are select(), explode(), map(), flatMap(), filter(), and where(), to mention some of them. Stateless operations support *append* and *update* output modes exclusively because information processed by these operations cannot be updated downstream. On the other hand, they do not support the *complete* output mode because it could be practically impossible to accumulate the complete stream of information.

In the next section, we are going to explore how to perform stateful operations and how the streaming configurations and resources have to be updated accordingly.

Spark Stateful Streaming

Stateful stream processing is stream processing that maintains events' states. It means that an event state is maintained and shared among events along the stream process; thus, event conditions can be maintained and/or updated over time. Stateful streams are used to persist live aggregates in streaming aggregations.

When it comes to stateful operations, Spark Structured Streaming provides a simple and concise API to maintain the state between different batches.

Stateful processing is necessary when we need to keep updated intermediate states (information being processed) along the streaming process, for example, when we need to perform data aggregation by key or event-time aggregations, assuming the ordered arrival time of events could not be guaranteed. Depicted in Figure 7-3, you can see how Spark uses the concept of *StateStore/state store* to maintain and share state information about events between various batches.

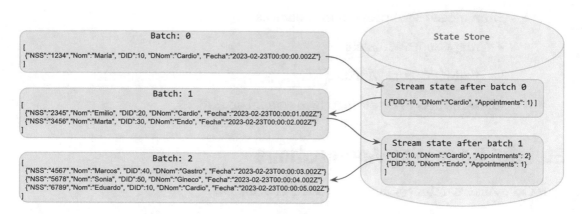

Figure 7-3. *Spark Structured Streaming state maintenance process*

Apache Spark achieves stateful operations by saving the intermediate state information in an in-memory key-value store named *StateStore*. The *StateStore* attains fault tolerance, saving intermediate states in a file system directory called a checkpoint directory.

Thus, with stateful streaming each micro-batch intermediate output is temporarily preserved and shared between executions using the event "state."

An example of stateful transformation is counting (`df.groupBy().count()`) the number of events processed since the beginning of a query. Spark keeps the number of events already counted in the event state and passes it to the next micro-batch, where that number is added to the ongoing count. Event state is maintained in the Spark executors' memory and saved to a designated file system directory to provide fault tolerance.

There are two types of stateful operations based on how intermediate information is managed and removed:

- Managed stateful operations are operations that automatically manage obsolete ("old") states. Operators such as

  - Streaming aggregations

  - Stream-stream joins

  - Streaming deduplication

  are part of this group.

- Unmanaged stateful operations, such as

  - mapGroupsWithState

  - flatMapGroupsWithState

  allow the user to define their own stateful operations.

Stateful Streaming Aggregations

Spark stateful streaming aggregations can be classified as

- *Time-based aggregations*: Number of events processed per unit of time

- *No-time-based aggregations*: Grouping events by key, for example

Time-Based Aggregations

Time-based aggregations are studied in detail in Chapter 8. Thus, we are going to leave it for now.

No-Time-Based Aggregations

No-time-based aggregations include

- Global aggregations

  Those are general aggregations with no key discrimination. In the examples you have seen so far, it could be the number of patients registering in a hospital:

```
# PySpark
counts = PatientDS.groupBy().count()
```

```
// Scala
val counts = PatientDS.groupBy().count()
```

- Grouped aggregations

  These are aggregations by key or with key discrimination. Adding to the previous Hospital Queue Management System application example, we could be interested in seeing only the number of appointments of specific medical departments in a hospital.

  In the following you can see a modified version of our Hospital Queue Management System application example with stateful grouped aggregations by department id and department name:

```
import org.apache.spark.sql.SparkSession
import org.apache.spark.sql.functions._
import org.apache.spark.sql.types.{IntegerType, StringType,
StructField, StructType,DoubleType,LongType}
import org.apache.spark.sql.{DataFrame, Dataset, Encoders,
SparkSession}
import java.io.IOException

val PatientsSchema = StructType(Array(
      StructField("NSS", StringType),
      StructField("Nom", StringType),
      StructField("DID", IntegerType),
      StructField("DNom", StringType),
      StructField("Fecha", StringType))
      )

case class Patient(
      NSS: String,
      Nom: String,
      DID: Option[Long],
      DNom: String,
      Fecha: String
)

val spark:SparkSession = SparkSession.builder()
      .master("local[*]")
      .appName("Hand-On-Spark3_Socket_Data_Source")
      .getOrCreate()
```

```scala
spark.sparkContext.setLogLevel("ERROR")

val host = "localhost"
val port = 9999

try {
    val PatientDS = spark.readStream
    .format("socket")
    .option("host",host)
    .option("port",port)
    .load()
    .select(from_json(col("value"),  PatientsSchema).
    as("patient"))
    .selectExpr("Patient.*")
    .as[Patient]

    printf("\n Listening and ready... \n")

    val counts = PatientDS
    .groupBy(col("DID"),col("DNom"))
    .count()

    counts.writeStream
    .format("update")
    .format("console")
    .outputMode("complete")
    .option("truncate",false)
    .option("newRows",30)
    .start()
    .awaitTermination()

} catch {
    case e: java.net.ConnectException => println("Error
    establishing connection to " + host + ":" + port)
    case e: IOException => println("IOException occurred")
    case t: Throwable => println("Error receiving data", t)
```

```
}finally {
      println("Finally block")
}
```

Running the previous code, you will see an output similar to the
next one:

```
Listening and ready...
-------------------------------------------
Batch: 1
-------------------------------------------
+---+------+-----+
|DID|DNom  |count|
+---+------+-----+
|20 |Neuro |4    |
|40 |Gastro|1    |
|50 |Gineco|3    |
|30 |Endo  |3    |
|10 |Cardio|9    |
+---+------+-----+

-------------------------------------------
Batch: 2
-------------------------------------------
+---+------+-----+
|DID|DNom  |count|
+---+------+-----+
|20 |Neuro |11   |
|40 |Gastro|3    |
|50 |Gineco|9    |
|30 |Endo  |8    |
|10 |Cardio|26   |
+---+------+-----+
```

```
--------------------------------------------
Batch: 3
--------------------------------------------
+---+------+-----+
|DID|DNom  |count|
+---+------+-----+
|20 |Neuro |11   |
|40 |Gastro|3    |
|50 |Gineco|9    |
|30 |Endo  |8    |
|10 |Cardio|27   |
+---+------+-----+
```

- Multiple aggregations

  The groupBy clause allows you to specify more than one aggregation function to transform column information. Therefore, multiple aggregations can be performed at once. For example, you can modify the previous code snippet as follows

```
val counts = PatientDS
.groupBy(col("DID"),col("DNom"))
.agg(count("*").alias("countDID"),
    sum("DID").alias("sumDID"),
    mean("DID").alias("meanDID"),
    stddev("DID").alias("stddevDID"),
    approx_count_distinct("DID").alias("distinctDID"),
    collect_list("DID").alias("collect_listDID"))
```

  to obtain several aggregations together.

  When you run the previous code and copy in your terminal (remember nc -lk 9999) first the following JSON strings

```
{"NSS":"4567","Nom":"Marcos", "DID":40, "DNom":"Gastro",
"Fecha":"2023-02-23T00:00:03.002Z"}
{"NSS":"5678","Nom":"Sonia", "DID":50, "DNom":"Gineco",
"Fecha":"2023-02-23T00:00:04.002Z"}
```

{"NSS":"6789","Nom":"Eduardo", "DID":10, "DNom":"Cardio",
"Fecha":"2023-02-23T00:00:05.002Z"}
{"NSS":"1001","Nom":"Lorena", "DID":10, "DNom":"Cardio",
"Fecha":"2023-02-23T00:00:06.002Z"}
{"NSS":"1006","Nom":"Sara", "DID":20, "DNom":"Neuro",
"Fecha":"2023-02-23T00:00:07.002Z"}
{"NSS":"1002","Nom":"Teresa", "DID":10, "DNom":"Cardio",
"Fecha":"2023-02-23T00:00:08.002Z"}
{"NSS":"1003","Nom":"Luis", "DID":20, "DNom":"Neuro",
"Fecha":"2023-02-23T00:00:09.002Z"}

and after that this second set of JSON strings

{"NSS":"1004","Nom":"Tomás", "DID":30, "DNom":"Endo",
"Fecha":"2023-02-23T00:00:10.002Z"}
{"NSS":"1005","Nom":"Lorena", "DID":50, "DNom":"Gineco",
"Fecha":"023-02-23T00:00:11.002Z"}
{"NSS":"1006","Nom":"Pedro", "DID":10, "DNom":"Cardio",
"Fecha":"023-02-23T00:00:12.002Z"}
{"NSS":"1007","Nom":"Ester", "DID":10, "DNom":"Cardio",
"Fecha":"023-02-23T00:00:13.002Z"}
{"NSS":"1008","Nom":"Marina", "DID":10, "DNom":"Cardio",
"Fecha":"023-02-23T00:00:14.002Z"}
{"NSS":"1009","Nom":"Julia", "DID":20, "DNom":"Neuro",
"Fecha":"023-02-23T00:00:15.002Z"}
{"NSS":"1010","Nom":"Javier", "DID":30, "DNom":"Endo",
"Fecha":"023-02-23T00:00:16.002Z"}
{"NSS":"1011","Nom":"Laura", "DID":50, "DNom":"Gineco",
"Fecha":"023-02-23T00:00:17.002Z"}
{"NSS":"1012","Nom":"Nuria", "DID":10, "DNom":"Cardio",
"Fecha":"023-02-23T00:00:18.002Z"}
{"NSS":"1013","Nom":"Helena", "DID":10, "DNom":"Cardio",
"Fecha":"023-02-23T00:00:19.002Z"}

you could see an output like the following one:

```
Listening and ready...
-------------------------------------------
Batch: 1
-------------------------------------------
+---+------+--------+------+-------+---------+-----------+--------------+
|DID|DNom  |countDID|sumDID|meanDID|stddevDID|distinctDID|collect_listDID|
+---+------+--------+------+-------+---------+-----------+--------------+
|20 |Neuro |2       |40    |20.0   |0.0      |1          |[20, 20]      |
|40 |Gastro|1       |40    |40.0   |null     |1          |[40]          |
|50 |Gineco|1       |50    |50.0   |null     |1          |[50]          |
|10 |Cardio|3       |30    |10.0   |0.0      |1          |[10, 10, 10]  |
+---+------+--------+------+-------+---------+-----------+--------------+


-------------------------------------------
Batch: 2
-------------------------------------------
+---+------+--------+------+-------+---------+-----------+--------------------------+
|DID|DNom  |countDID|sumDID|meanDID|stddevDID|distinctDID|collect_listDID           |
+---+------+--------+------+-------+---------+-----------+--------------------------+
|20 |Neuro |3       |60    |20.0   |0.0      |1          |[20, 20, 20]              |
|40 |Gastro|1       |40    |40.0   |null     |1          |[40]                      |
|50 |Gineco|3       |150   |50.0   |0.0      |1          |[50, 50, 50]              |
|30 |Endo  |2       |60    |30.0   |0.0      |1          |[30, 30]                  |
|10 |Cardio|7       |70    |10.0   |0.0      |1          |[10, 10, 10, 10, 10, 10, 10]|
+---+------+--------+------+-------+---------+-----------+--------------------------+


-------------------------------------------
Batch: 3
-------------------------------------------
```

```
+---+------+--------+------+-------+---------+----------+----------------------------------+
|DID|DNom  |countDID|sumDID|meanDID|stddevDID|distinctDID|collect_listDID                  |
+---+------+--------+------+-------+---------+----------+----------------------------------+
|20 |Neuro |3       |60    |20.0   |0.0      |1         |[20, 20, 20]                      |
|40 |Gastro|1       |40    |40.0   |null     |1         |[40]                              |
|50 |Gineco|3       |150   |50.0   |0.0      |1         |[50, 50, 50]                      |
|30 |Endo  |2       |60    |30.0   |0.0      |1         |[30, 30]                          |
|10 |Cardio|8       |80    |10.0   |0.0      |1         |[10, 10, 10, 10, 10, 10, 10, 10]  |
+---+------+--------+------+-------+---------+----------+----------------------------------+
```

Note   The aggregation functions shown in the previous code snippet are included for illustration purposes only. Obviously, functions such as sum(), mean(), stddev(), approx_count_distinct(), and collect_list() applied to a medical department id "DID" do not make any business sense.

- Built-in aggregation functions

  Spark Streaming built-in aggregation functions simplify the process of summarizing data, which is an important component of data analytics. To use them, you need to specify an aggregation key for grouping and the aggregation function that defines how the transformations will be performed across the DataFrame columns.

  Table 7-3 shows a list of the most common aggregation functions for DataFrames. A complete list of aggregation functions for column operations can be found in the official documentation.[6]

---

[6]https://spark.apache.org/docs/3.3.2/api/R/reference/column_aggregate_functions.html

Table 7-3. *Spark Structured Streaming List of Aggregation Functions for Dataframes*

Aggregation Function	Description
approx_count_distinct()	Returns a new column for approximate distinct count of a column.
avg()	Returns the average of the values in a group.
collect_list()	Returns a list with duplicates of all values from an input column.
collect_set()	Returns a set of the values from an input column without duplicates.
countDistinct()(*)	Returns a new column for distinct elements in a column.
count()	Returns the number of elements in a column.
first()	Returns the first non-null element in a column.
last()	Returns the last non-null element.
kurtosis()	Returns the kurtosis of the values in a column. It could be used to try to identify outliers in the data.
max()/min()	They return the maximum or minimum value in a column.
mean()	An alias for avg(), it returns the average of the elements in a column.
skewness()	Returns the skewness of the values in a column. It is the degree of distortion from the normal distribution.
stddev()	An alias for stddev_samp(), it returns the unbiased sample standard deviation of the expression in a group.
stddev_pop()	Returns population standard deviation of the expression in a column.

(continued)

Table 7-3. (*continued*)

Aggregation Function	Description
sum()	Returns the sum of the values in a column.
sumDistinct()	Returns the sum of the distinct values in a column.
variance()	An alias for var_samp(), it returns the unbiased sample variance of the values in a column.
var_pop()	Returns the population variance of the values in a column.

(*) Although countDistinct() appears in the literature available as a valid aggregation function for Structured Streaming, at the time this book was written, the following message was outputted by Spark when we tried to use it as one of the multiple aggregation functions described in the previous section:

```
"Distinct aggregations are not supported on streaming DataFrames/Datasets.
Consider using approx_count_distinct() instead."
```

- User-defined aggregation

  Finally, Spark Structured Streaming supports user-defined aggregation functions. Check the Spark SQL Guide for more and updated details.

7.6 Spark Checkpointing Streaming

To provide fault tolerance, Spark uses checkpointing to ensure it can recover from failures. Checkpointing is used to persist intermediate information states in a file system storage from which Spark can read upon failure. In stateful streaming, it is mandatory to apply checkpointing to be able to restore transitional states in the eventuality of a failure.

The *StateStore* studied in the "Spark Stateful Streaming" section and depicted in Figure 7-3 supports incremental checkpointing, meaning that only the key-values updated are preserved, without modifying other key-value pairs present in the streaming process.

To include checkpointing support in our streaming Hospital Queue Management System application example, we are going to update our previous code snippet as follows:

```
import org.apache.spark.sql.streaming._
// ...
val checkpointDir = "/tmp/streaming_checkpoint"
// ...
counts.writeStream
    // ...
    .trigger(Trigger.ProcessingTime("5 seconds"))
    .option("checkpointLocation", checkpointDir)
    // ...
    .start()
    .awaitTermination()
```

As you can see, we have introduced several new features that we explain in the following:

- *Trigger*: Defines how often a streaming query must be triggered (run) to process newly available streaming data—in other words, how frequently our application has to review the data sources looking for new information and possibly emit new data. Trigger was introduced into Spark to set the stream batch period.

  ProcessingTime is a trigger that assumes milliseconds as the minimum unit of time. ProcessingTime(interval: String) accepts CalendarInterval instances with or without interval strings, for example:

  - *With interval strings*: ProcessingTime("interval 10 seconds")

  - *Without interval strings*: ProcessingTime("10 seconds")

  There are four factory methods (options):

  - *Default*: If no trigger is set, the streaming query runs micro-batches one after another, as soon as the precedent micro-batch has finished.

- *OneTimeTrigger*: With this trigger mode set, it executes the trigger once and stops. The streaming query will execute the data available in only one micro-batch. A use case for this trigger mode could be to use it as a kind of daily batch processing, saving computing resources and money. Example: `.trigger(Trigger.Once)`.

- *ProcessingTime*: The user can define the ProcessingTime parameter, and the streaming query will be triggered with the interval established, executing new micro-batches and possibly emitting new data.

- *ContinuousTrigger*: At the time this book was written, continuous processing was an experimental streaming execution mode introduced in Spark 2.3.[7] It has been designed to achieve low latencies (in the order of 1 ms) providing at-least-once guarantee. To provide fault tolerance, a checkpoint interval must be provided as a parameter. Example: `.trigger(Trigger.Continuous("1 second"))`. A checkpoint interval of 1 s means that the stream engine will register the intermediate results of the query every second. Every checkpoint is written in a micro-batch engine-compatible structure; therefore, after a failure, the ongoing (supported) query can be restarted by any other kind of trigger. For example, a supported query that was started using the micro-batch mode can be restarted in continuous mode, and vice versa. The continuous processing mode only supports stateless queries such as select, map, flatMap, mapPartitions, etc. and selections like where, filter, etc. All the SQL functions are supported in continuous mode, except aggregation functions `current_timestamp()` and `current_date()`.

- checkpointLocation

---

[7] For up-to-date information, please check the Apache Spark official documentation, `https://spark.apache.org/docs/latest/structured-streaming-programming-guide.html#continuous-processing`

This parameter points to the file system directory created for state storage persistence purposes. To make the store fault-tolerant, the option checkpointLocation must be set as part of the writeStream output configuration.

The state storage uses the checkpoint folder to store mainly

- Data checkpointing

- Metadata checkpointing

In case we are using stateful operations, the structure of the Spark Streaming checkpoint folder and the state data representation folders will look as illustrated in Table 7-4.

Table 7-4. *Spark Streaming Checkpoint and the State Data Representation Structure*

The **checkpointLocation** Folder Structure	The State Data Representation
/tmp/streaming_checkpoint	└── state
├── commits	└── 0
│ ├── 0	├── 0
│ ├── 1	│ ├── 1.delta
│ ├── 2	│ ├── 2.delta
│ └── 3	│ ├── 3.delta
├── metadata	│ ├── 4.delta
├── offsets	│ └── _metadata
│ ├── 0	│ └── schema
│ ├── 1	├── 1
│ ├── 2	│ ├── 1.delta
│ └── 3	│ ├── 2.delta
└── state	│ ├── 3.delta
└── 0	│ └── 4.delta
	├── 10
	│ ├── 1.delta
	│ ├── 2.delta
	│ ├── 3.delta
	│ └── 4.delta

Recovering from Failures with Checkpointing

Upon failure or intentional shutdown, the intermediate information persisted inside the checkpoint directory can be used to restore the query exactly where it stopped. In addition, after restoration some changes are allowed in a streaming query and some others not. For example, you can change the query sink from file to Kafka but not vice versa. You can check out the most updated list of allowed and not allowed changes in a streaming query between restarts from the same checkpoint location, looking at the Structured Streaming Programming Guide documentation.[8]

7.7 Summary

In this chapter we went over the Spark Structured Streaming module. Firstly, we studied the general semantics of message delivery reliability mechanisms. Secondly, we compared Structured Streaming with Spark Streaming based on DStreams. After that, we explained the technical details behind the Spark Structured Streaming architecture, such as input and result tables as well as the different output modes supported. In addition, we also went through the streaming API for DataFrames and datasets and Structured Streaming stateless and stateful transformations and aggregations, giving some interesting examples that will help you learn how to implement these features. Finally, we studied the concepts of streaming checkpointing and recovery, giving some practical examples. In the next chapter, we are moving forward studying streaming sources and sinks.

---

[8] https://spark.apache.org/docs/3.3.2/structured-streaming-programming-guide. html#recovery-semantics-after-changes-in-a-streaming-query

CHAPTER 8

Streaming Sources and Sinks

In the previous chapter, we went through the basics of an end-to-end Structured Streaming process. Remember the foundations of Apache Spark Structured Streaming are creating streaming dataframes by ingesting data from a source using the `SparkSession.readStream()` method, applying business logic to it using the processing engine and outputting the result DataFrame to a data sink using `DataFrame.writeStream()`.

In this chapter we are going to delve into the usage of built-in data sources and sinks, as well as how to create your own custom streaming sources and sinks using `foreachBatch()` and `foreach()` methods to implement your own functionality and write your data to a storage system other than that natively supported by Spark Structured Streaming.

8.1 Spark Streaming Data Sources

Remember we saw in the previous chapter Spark supports various input sources for data ingestion. Some of them are the so-called built-in sources:

- *File source*: It is used for streaming data from a file system. Supported file formats are text, CSV, JSON, and Parquet.

- *Kafka source*: It is used for reading data from Kafka topics. It requires Kafka version 0.10.0 or higher.

Then, there are other data sources considered mostly for testing as they are not fault-tolerant:

- *Socket source*: It reads the data from a TCP/IP socket connection.

- *Rate source*: It generates random data at the specified number of rows per second, where each row of data has two columns: a "timestamp" and a "value."

© Alfonso Antolínez García 2023
A. Antolínez García, *Hands-on Guide to Apache Spark 3*, https://doi.org/10.1007/978-1-4842-9380-5_8

- *Rate per micro-batch source*: It is similar to the "rate" source, but in this case it produces a consistent set of input rows per micro-batch regardless of query execution configuration. For example, batch 0 will produce values in the interval [0, 999], batch 1 will generate values in the interval [1000, 9999], and so on.

Reading Streaming Data from File Data Sources

Apache Spark Structured Streaming natively supports stream reading from file systems employing the same file formats as those supported in batch processing (text, CSV, JSON, ORC, and Parquet).

Spark Structured Streaming uses the DataStreamReader class for streaming text files from a file system folder. When you define a directory as a streaming source, Spark treats the files appearing in that location as a data stream. That means a FileStreamSource is a source that reads text format files from a directory as they are seen by Spark. Next is an example of how to set up a basic file source streaming:

```
val df = spark.readStream
  .format("text")
  .option("maxFilesPerTrigger", 1)
  .load("/tmp/logs")
```

You can also specify the schema of your data, for example:

```
val PatientsSchema = StructType(Array(
    StructField("NSS", StringType),
    StructField("Nom", StringType),
    StructField("DID", IntegerType),
    StructField("DNom", StringType),
    StructField("Fecha", StringType) )
  )
```

And then you can read the files based on the precedent schema:

```
val df = spark
    .readStream
    .schema(PatientsSchema)
    .json("/tmp/patient_streaming")
```

In our preceding example, the returned df streaming DataFrame will have the PatientsSchema. The "*/tmp/patient_streaming*" source directory must exist when the stream process starts.

There are some important points to remember when using file sources:

- The source directory must exist when the stream process starts, as mentioned before.

- All the files streamed to the source directory must be of the same format, that is to say, they all must be text, JSON, Parquet, etc., and the schema must be also the same if we want to preserve data integrity and avoid errors.

- Files already present in the designated folder when the streaming job begins are ignored. This concept is depicted in Figure 8-1.

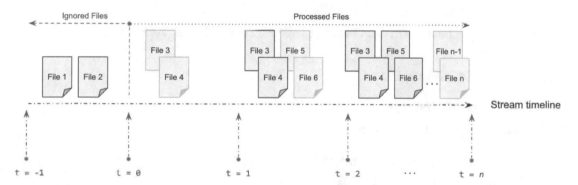

Figure 8-1. *File data stream processing schema*

- Spark uses system tools that list files to identify the new files. Therefore, the files appearing in the streaming directory must be complete and closed, because Spark will process them as soon as they are discovered. Thus, any data addition or file update could result in data loss.

- When Spark processes a file, it is internally labeled as processed. Hence, it will not be processed again even if it is updated.

- In case several files should be processed, but Spark can only cope with part of them in the next micro-batch, files with the earliest timestamps will be processed first.

When creating a new FileStreamSource instance, two main options are available:

- *schema*: As we have already mentioned, it is the schema of the data, and it is specified at instantiation time.

- *maxFilesPerTrigger*: It specifies the maximum number of files read per micro-batch. Therefore, it is used to control the stream read rate to the maximum number of files per trigger.

In the following you have a code example in which we stream data from a file source. This example includes the schema of the files used as a data source, streams data from a directory, and outputs the results of the transformation to the console:

```
package org.apress.handsOnSpark3

import org.apache.spark.sql.SparkSession
import org.apache.spark.sql.functions._
import org.apache.spark.sql.types.{IntegerType, StringType, StructField,
StructType}
import java.io.IOException

object dStreamsFiles {
  def main(args: Array[String]): Unit = {

    val PatientsSchema = StructType(Array(
    StructField("NSS", StringType),
    StructField("Nom", StringType),
    StructField("DID", IntegerType),
    StructField("DNom", StringType),
    StructField("Fecha", StringType)
    )
    )

    try {
    val spark: SparkSession = SparkSession
    .builder()
    .master("local[3]")
    .appName("Hand-On-Spark3_File_Data_Source")
    .getOrCreate()
```

```
spark.sparkContext.setLogLevel("ERROR")

val df = spark
.readStream
.schema(PatientsSchema)
.json("/tmp/patient_streaming")

val groupDF = df.select("DID")
.groupBy("DID").agg(count("DID").as("Accumulated"))
.sort(desc("Accumulated"))

printf("\n Listening and ready... \n")

groupDF.writeStream
.format("console")
.outputMode("complete")
.option("truncate", false)
.option("newRows", 30)
.start()
.awaitTermination()

} catch {
case e: IOException => println("IOException occurred")
case t: Ihrowable => println("Error receiving data", t)
} finally {
println("Finally block")
}

 }
}
```

Next, we are going to jump to another built-in data source and one of the most commonly used nowadays. First, we are going to provide an introduction about Kafka, and after that we are going to provide a practical example.

Reading Streaming Data from Kafka

Apache Kafka is an open source, distributed, persistent, and highly scalable event streaming platform enabling the development of real-time, event-driven applications, among other features. Kafka organizes and stores events in topics, which are Kafka's most fundamental unit of organization. A topic is a named log of events, something similar to a table in a relational database. Events in the log are immutable and durable, meaning that once something has happened, they cannot be altered or deleted—they persist unchanged and can remain in the topic for a defined period of time or, eventually, indefinitely. Kafka logs/topics containing the events are files stored on a disk.

An example of how Kafka components interact in the Kafka architecture is shown in Figure 8-2.

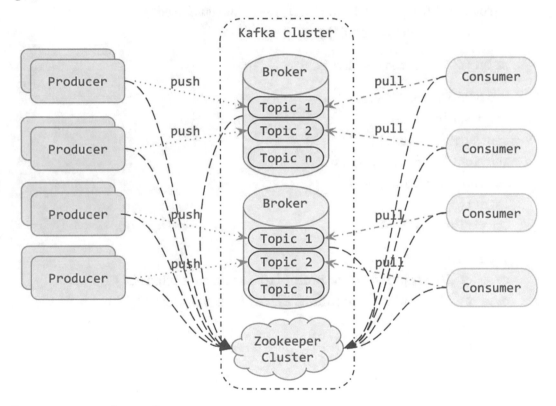

Figure 8-2. *Kafka architecture*

Kafka has four primary capabilities:

- Kafka allows applications to publish or subscribe to event streams, enabling them to respond to events in real time.

- Kafka manages records preserving the order in which they occurred.

- Kafka is a fault-tolerant and scalable system that processes records in real time.

- The simple semantics of topics allow Kafka to deliver high levels of sustained in and out throughput and facilitate data replication to enhance fault tolerance. Kafka topics are partitioned and replicated, contributing to maintaining a high-performance simultaneous event delivering service to a large number of consumers.

In Figure 8-3 you can see a graphical representation of the Kafka concepts of topic, data partition, and replica.

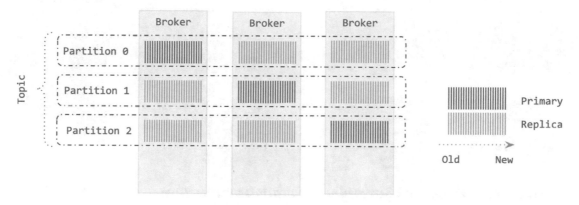

Figure 8-3. *Kafka concepts of topic, partition, and replica*

Kafka capabilities can be leveraged through four APIs:

- *Producer API*: It is used to publish an event or stream of events to a Kafka topic.

- *Consumer API*: Applications use it to subscribe to one or more topics consuming the stream of data stored in the topic. Information in a topic can be consumed in real time or can be read from historical registers.

- *Streams API*: This API is more complex than the Producer and Consumer APIs and provides the capacity to build complex data and event streaming processes. For example, it can be used to set up end-to-end stream jobs, receiving information, analyzing it, and transforming it, if required.

- *Connector API*: This API is intended for the development of connectors, to automate the data flow to and from a Kafka cluster.

Apache Kafka and Spark Streaming are often used together to process real-time data streams. Coupling Kafka and Spark can lead to a reliable, performant, and scalable streaming data processing pipeline able to cope with complex event processes. A sketch of the Kafka-Spark Streaming integration architecture is depicted in Figure 8-4.

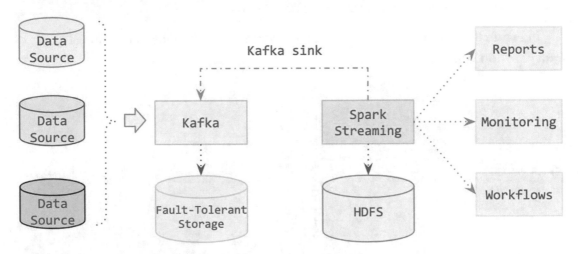

Figure 8-4. *Kafka-Spark Streaming integration architecture*

Implementing a data processing pipeline using Kafka-Spark Streaming includes data intake from Kafka topics, manipulating and analyzing data using Spark Streaming, and then storing the treated data in a final sink or injecting it back again into another Kafka topic as part of another pipeline.

In the following code snippet, you have an example of how Apache Kafka and Apache Spark Structured Streaming can work together to implement a highly scalable real-time processing architecture:

```
package org.apress.handsOnSpark3.com

import org.apache.spark.sql.SparkSession
import org.apache.spark.sql.functions.{col, from_json}
import org.apache.spark.sql.types.{IntegerType, StringType, StructType,
StructField}

object SparkKafka {

  def main(args: Array[String]): Unit = {

      val spark: SparkSession = SparkSession.builder
```

```scala
        .master("local[3]")
        .appName("SparkStructuredStreamingHospital")
        .getOrCreate()

    spark.sparkContext.setLogLevel("ERROR")

    import spark.implicits._

    val df = spark.readStream
    .format("kafka")
    .option("kafka.bootstrap.servers", "localhost:9092")
    .option("subscribe", "patient")
    .option("startingOffsets", "earliest")
    .load()

    df.printSchema()

    val PatientsSchema = StructType(Array(
    StructField("NSS", StringType),
    StructField("Nom", StringType),
    StructField("DID", IntegerType),
    StructField("DNom", StringType),
    StructField("Fecha", StringType))
    )

    val patient = df.selectExpr("CAST(value AS STRING)")
    .select(from_json(col("value"), PatientsSchema).as("data"))
    .select("data.*")

    patient.printSchema()

    val query = patient.writeStream
    .format("console")
    .outputMode("append")
    .start()
    .awaitTermination()
  }
}
```

As soon as you have your code ready, it is time to give it a try. The first thing we are going to do is to start the Kafka environment.

Note   At the time this book was written, Kafka 3.4.0 was the latest release and the one used in our examples. To be able to execute the code shown before, your local environment must have Java 8+ installed.

Apache Kafka can be started using ZooKeeper or KRaft. In this book we are using only the former.

Firstly, open a terminal session and from your $KAFKA_HOME directory execute the following commands in order to start all services in the correct order. Run the following commands to start the ZooKeeper service with the default configuration:

```
$ bin/zookeeper-server-start.sh config/zookeeper.properties
```

Secondly, open another terminal session and run the following commands to start the Kafka broker service with the default configuration as well:

```
$ bin/kafka-server-start.sh config/server.properties
```

As soon as all preceding services are successfully running, a basic Kafka environment will be ready to use. However, before we can write our first events, we have to create a topic.

Therefore, open another terminal session and run the following code to create a "patient" topic to use with our Hospital Queue Management System data examples:

```
$ bin/kafka-topics.sh --create --topic patient --bootstrap-server
localhost:9092
```

The kafka-topics.sh command without any arguments can also be used to display usage information. For example, it can be employed to show the details of the new topic, such as the partition count, replicas, etc. of the patient topic. You can execute the following command and options, to display that information:

```
bin/kafka-topics.sh --describe --topic patient --bootstrap-server
localhost:9092
```

```
Topic: patient     TopicId: Bhq8M7cgTVqRrV18nT7dzg     PartitionCount:
1     ReplicationFactor: 1     Configs:
    Topic: patient     Partition: 0     Leader: 0     Replicas: 0     Isr: 0
```

Now is time to write some events into the *patient* topic just created and see the results. To do that, we are going to create a Kafka producer using the "bin/kafka-console-producer.sh", which is located in the Kafka directory.

A Kafka producer is a client application that communicates with the Kafka brokers for writing events into topics. Once the information is received, the brokers will save it in a fault-tolerant storage for as long as we could need it, allegedly forever. This is the reason our Spark application is going to be able to asynchronously consume the information stored in our example topic.

To see how it works, open a new terminal session and run the producer console client, as shown in the following, to write some events into our "patient" topic just created. In this example we are going to use the data from the JSON files of Chapter 6. By default, every line you type will be a new event being written to the "patient" topic:

```
$ bin/kafka-console-producer.sh --topic patient --bootstrap-server
localhost:9092
>{"NSS":"4567","Nom":"Marcos", "DID":40, "DNom":"Gastro", "Fecha":"01-09-2022"}
>{"NSS":"5678","Nom":"Sonia", "DID":50, "DNom":"Gineco", "Fecha":"01-09-2022"}
>{"NSS":"6789","Nom":"Eduardo", "DID":10, "DNom":"Cardio",
"Fecha":"01-09-2022"}
>{"NSS":"1234","Nom":"María", "DID":10, "DNom":"Cardio", "Fecha":"01-09-2022"}
>{"NSS":"4567","Nom":"Marcos", "DID":40, "DNom":"Gastro", "Fecha":"01-09-2022"}
> . . .
> . . .
> . . .
>{"NSS":"2345","Nom":"Emilio", "DID":20, "DNom":"Neuro", "Fecha":"01-09-2022"}
>{"NSS":"3456","Nom":"Marta", "DID":30, "DNom":"Endo", "Fecha":"01-09-2022"}
>{"NSS":"4567","Nom":"Marcos", "DID":40, "DNom":"Gastro",
"Fecha":"01-09-2022"}
>{"NSS":"4567","Nom":"Marcos", "DID":40, "DNom":"Gastro",
"Fecha":"01-09-2022"}
>{"NSS":"5678","Nom":"Sonia", "DID":50, "DNom":"Gineco", "Fecha":"01-09-2022"}
>{"NSS":"6789","Nom":"Eduardo", "DID":10, "DNom":"Cardio",
"Fecha":"01-09-2022"}
>{"NSS":"1234","Nom":"María", "DID":10, "DNom":"Cardio",
"Fecha":"01-09-2022"}
```

After pasting the content of the JSON files onto the Kafka producer console, run your example program as follows:

```
$SPARK_HOME/bin/spark-submit --class org.apress.handsOnSpark3.com.
SparkKafka --master yarn --packages org.apache.spark:spark-sql-
kafka-0-10_2.12:3.2.0 /PATH/TO/JAR/FILE/HandsOnSpark3-Structured_Streaming_
Hospital-1.0.jar
```

As soon as the program is running, you could see an output similar to the next one coming out from your program:

```
root
 |-- key: binary (nullable = true)
 |-- value: binary (nullable = true)
 |-- topic: string (nullable = true)
 |-- partition: integer (nullable = true)
 |-- offset: long (nullable = true)
 |-- timestamp: timestamp (nullable = true)
 |-- timestampType: integer (nullable = true)

root
 |-- NSS: string (nullable = true)
 |-- Nom: string (nullable = true)
 |-- DID: integer (nullable = true)
 |-- DNom: string (nullable = true)
 |-- Fecha: string (nullable = true)

-------------------------------------------
Batch: 0
-------------------------------------------
+----+-----+---+------+----------+
| NSS|  Nom|DID|  DNom|     Fecha|
+----+-----+---+------+----------+
|1234|María| 10|Cardio|01-09-2022|
+----+-----+---+------+----------+
```

```
-------------------------------------------
Batch: 1
-------------------------------------------
+----+------+---+------+----------+
| NSS|   Nom|DID| DNom|     Fecha|
+----+------+---+------+----------+
|4567|Marcos|  40|Gastro|01-09-2022|
|5678| Sonia|  50|Gineco|01-09-2022|
+----+------+---+------+----------+

-------------------------------------------
Batch: 2
-------------------------------------------
+----+-------+---+------+----------+
| NSS|    Nom|DID| DNom|     Fecha|
+----+-------+---+------+----------+
|6789|Eduardo|  10|Cardio|01-09-2022|
+----+-------+---+------+----------+

-------------------------------------------
Batch: 3
-------------------------------------------
+----+------+---+------+----------+
| NSS|   Nom|DID| DNom|     Fecha|
+----+------+---+------+----------+
|1234| María|  10|Cardio|01-09-2022|
|2345|Emilio|  20| Neuro|01-09-2022|
|3456| Marta|  30|  Endo|01-09-2022|
|4567|Marcos|  40|Gastro|01-09-2022|
+----+------+---+------+----------+
```

```
--------------------------------------------
Batch: 4
--------------------------------------------
+----+-------+---+------+----------+
| NSS|    Nom|DID| DNom|     Fecha|
+----+-------+---+------+----------+
|4567| Marcos| 40|Gastro|01-09-2022|
|5678|  Sonia| 50|Gineco|01-09-2022|
|6789|Eduardo| 10|Cardio|01-09-2022|
|1234|  María| 10|Cardio|01-09-2022|
|4567| Marcos| 40|Gastro|01-09-2022|
|5678|  Sonia| 50|Gineco|01-09-2022|
|6789|Eduardo| 10|Cardio|01-09-2022|
|1234|  María| 10|Cardio|01-09-2022|
|2345| Emilio| 20| Neuro|01-09-2022|
|3456|  Marta| 30|  Endo|01-09-2022|
|4567| Marcos| 40|Gastro|01-09-2022|
+----+-------+---+------+----------+
```

To double-check the results of your streaming process, you can also read the events from the Kafka brokers using a Kafka consumer, which is a client application that subscribes to (reads and processes) events.

To see how that works, open another terminal session and run the consumer console client as shown in the following, to read the *patient* topic we created before:

```
$ bin/kafka-console-consumer.sh --topic patient --from-beginning
--bootstrap-server localhost:9092
```

You will see on your screen an output similar to the following one:

```
{"NSS":"1234","Nom":"María", "DID":10, "DNom":"Cardio", "Fecha":"01-09-2022"}
{"NSS":"2345","Nom":"Emilio", "DID":20, "DNom":"Neuro", "Fecha":"01-09-2022"}
{"NSS":"3456","Nom":"Marta", "DID":30, "DNom":"Endo", "Fecha":"01-09-2022"}
{"NSS":"4567","Nom":"Marcos", "DID":40, "DNom":"Gastro",
"Fecha":"01-09-2022"}
. . .
. . .
```

```
{"NSS":"4567","Nom":"Marcos", "DID":40, "DNom":"Gastro",
"Fecha":"01-09-2022"}
{"NSS":"5678","Nom":"Sonia", "DID":50, "DNom":"Gineco", "Fecha":"01-09-2022"}
{"NSS":"6789","Nom":"Eduardo", "DID":10, "DNom":"Cardio",
"Fecha":"01-09-2022"}
```

To be able to compile the code examples used in this section, you have to use the correct Kafka dependencies and Scala compiler version, and all depend on your Kafka, Spark, and Scala versions installed.

So far, we have talked about Spark built-in streaming data sources like TCP/IP sockets, files, Apache Kafka, etc. Other advanced streaming applications that can be paired with Apache Spark to create stream pipes could be Kinesis. In the next section, we are going to see how to create custom stream data sources using tools primarily not intended for that purpose. In particular we are going to show you how to stream data from a NoSQL database such as MongoDB.

Reading Streaming Data from MongoDB

Spark Streaming allows live analysis of data streams read from MongoDB. In this section we are going to stream data between MongoDB and Spark using Spark Structured Streaming and the new continuous processing trigger.

To accomplish our task, we are also going to use the new v2 MongoDB Spark connector. The latest 10.x series connector provides native integration between Spark Structured Streaming and MongoDB and supports the new continuous trigger-type streaming. This connector also takes advantage of one of the features of MongoDB version 5.1 and onward, called a *"change stream cursor,"* to subscribe to information changes in the database.

Therefore, with this connector, we are going to open an input stream connection from our MongoDB database and at the same time set up a MongoDB change stream cursor to the designated database and data collection. This feature triggers a change stream event as soon as new documents are inserted or the existing ones are modified or deleted. Those event changes are forwarded to the specified consumer, in our case Spark.

In Figure 8-5 you can see an example use case in which Spark and MongoDB are coupled to build an event streaming architecture.

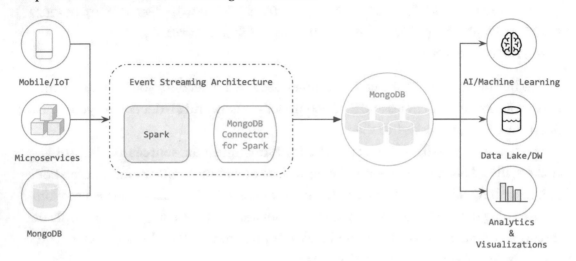

Figure 8-5. *Spark-MongoDB event streaming architecture*

Consider the following example that streams live information regarding medical appointments from a MongoDB Atlas cluster onto our Hospital Queue Management System application we have been using in previous chapters:

```
{"NSS":"2345","Nom":"Emilio", "DID":20, "DNom":"Neuro",
"Fecha":"01-09-2022"}
{"NSS":"3456","Nom":"Marta", "DID":30, "DNom":"Endo", "Fecha":"01-09-2022"}
```

Information like this in a MongoDB document looks as follows:

```
{
  "_id": {
      "$oid": "640cba70f9972564d8c4ef2f"
  },
  "NSS": "2345",
  "Nom": "Emilio",
  "DID": 20,
  "DNom": "Neuro",
  "Fecha": "01-09-2022"
}
```

In the next code snippet, we will use the new MongoDB Spark connector to read data from our MongoDB data collection:

```
import org.apache.spark.sql.SparkSession
import org.apache.spark.{SparkConf, SparkContext}
import org.apache.spark.sql.streaming.Trigger

val spark:SparkSession = SparkSession
      .builder()
      .config("spark.jars.packages", "org.mongodb.spark:mongo-spark-
      connector:10.1.1")
      .master("local[*]")
      .appName("Hand-On-Spark3_File_Data_Source_MongoDB")
      .getOrCreate()

val sc = spark.sparkContext

sc.setLogLevel("ERROR")

val mongoDBURI = "mongodb+srv://<user>:<password>@hands-on-spark3.abcdef.
mongodb.net/?retryWrites=true&w=majority"

val columsOfInterest = List("NSS","Nom","DID","DNom","Fecha","_id")

// define a streaming query
val df = spark.readStream
  .format("mongodb")
  .option("spark.mongodb.connection.uri", mongoDBURI)
  .option("spark.mongodb.database", "MongoDB_Data_Source")
  .option("spark.mongodb.collection", "MongoDB_Data_Source")
  .option("spark.mongodb.change.stream.publish.full.document.only", "true")
  .option("forceDeleteTempCheckpointLocation", "true")
  //.schema(PatientsSchema)
  .load()

df.printSchema()

if (df.isStreaming) printf(" ----- Streaming is running -----! \n")

import spark.implicits._
```

```
val groupDF = df.select(columsOfInterest.map(col): _*) // Here you could do
data transformation
```

```
groupDF.printSchema()
```

```
groupDF.writeStream
      .outputMode("append")
      .option("forceDeleteTempCheckpointLocation", "true")
      .format("console")
      .option("checkpointLocation", "/tmp/checkpointDir")
      //.trigger(Trigger.ProcessingTime("10 seconds"))
      .trigger(Trigger.Continuous("30 seconds"))
      .start()
      .awaitTermination()
```

Going through the preceding code, you notice that while reading from a MongoDB database, we do not necessarily need to define an information schema as the schema is inferred from the MongoDB collection.

In any case, if you prefer or need to define your data schema, you can do it and call the stream read process as follows.

First, define the schema of your data:

```
val PatientsSchema = StructType(Array(
     StructField("NSS", StringType),
     StructField("Nom", StringType),
     StructField("DID", IntegerType),
     StructField("DNom", StringType),
     StructField("Fecha", StringType),
   StructField("_id", StringType))
     )
```

After that, use it in combination with your readStream method like this, to define the schema of the incoming data:

```
val df = spark.readStream
  .format("mongodb")
  .option("spark.mongodb.connection.uri", mongoDBURI)
  .option("spark.mongodb.database", "MongoDB_Data_Source")
```

```
.option("spark.mongodb.collection", "MongoDB_Data_Source")
.option("spark.mongodb.change.stream.publish.full.document.only", "true")
.option("forceDeleteTempCheckpointLocation", "true")
.schema(PatientsSchema)
.load()
```

Another code line to pay attention to is the following one:

```
if (df.isStreaming) printf(" ----- Streaming is running -----! \n")
```

We have used the property *isStreaming* to verify that the dataset is streaming. It returns true if the df dataset contains one or more data sources that constantly send data as it arrives.

Finally, when writing the streamed data to the console, we have chosen the continuous trigger type as it is supported by the latest MongoDB Spark connector.

In this case, we have set the trigger to "30 s" for the sake of readability, as using a "1 s" trigger, for instance, would have been pulling data continuously to the console and it would have been more difficult to collect it for the book:

```
.trigger(Trigger.Continuous("30 seconds"))
```

Nevertheless, you can use any of the other supported trigger types, such as

- *Default trigger*: It runs micro-batches as soon as possible.

- *ProcessingTime trigger*: It triggers micro-batches with a time interval specified.

- *One-time trigger*: It will execute only one micro-batch, process the information available, and stop.

- *Available-now trigger*: It is similar to the one-time trigger with the difference that it is designed to achieve better query scalability trying to process data in multiple micro-batches based on the configured source options (e.g., *maxFilesPerTrigger*).

In the next code example, we show how to modify the previous program to use Trigger.ProcessingTime with a 10 s interval:

```
groupDF.writeStream
    .outputMode("append")
    .option("forceDeleteTempCheckpointLocation", "true")
```

```
.format("console")
.option("checkpointLocation", "/tmp/checkpointDir")
.trigger(Trigger.ProcessingTime("10 seconds"))
.start()
.awaitTermination()
```

Well, now it is time to give the program a try.

As soon as you execute the program, insert some documents (information) in your database. You can do as it as displayed in Figure 8-6 if you are using a graphical interface such as MongoDB Compass.

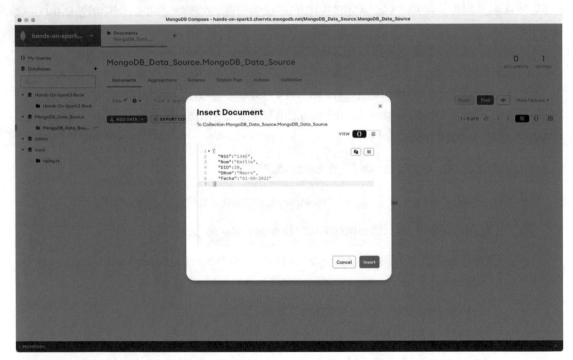

***Figure 8-6.** Inserting a new document into the MongoDB database*

Once the new document is inserted into the MongoDB database, you can see it displayed as in Figure 8-7.

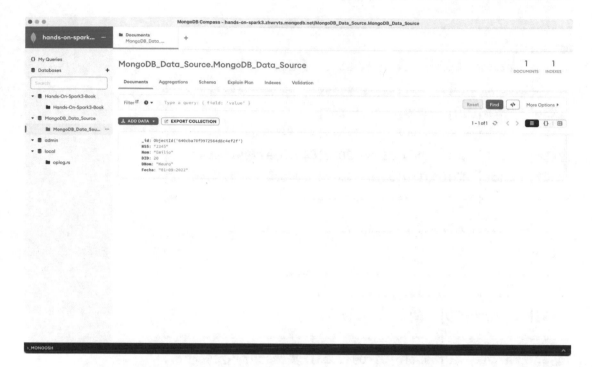

Figure 8-7. *A new document is inserted into a MongoDB database and collection*

You should see an outcome similar to the following one coming out from your application:

```
root
|-- _id: string (nullable = true)
|-- NSS: string (nullable = true)
|-- Nom: string (nullable = true)
|-- DID: integer (nullable = true)
|-- DNom: string (nullable = true)
|-- Fecha: string (nullable = true)

----- Streaming is running -----!

… removed for brevity …
```

```
-------------------------------------------
Batch: 1
-------------------------------------------
+----+------+---+------+----------+-------------------+
| NSS|   Nom|DID|  DNom|     Fecha|                _id|
+----+------+---+------+----------+-------------------+
|3456| Marta| 30|  Endo|01-09-2022|640cbaa7f9972564d...|
|4567|Marcos| 40|Gastro|01-09-2022|640cbab3f9972564d...|
+----+------+---+------+----------+-------------------+


-------------------------------------------
Batch: 2
-------------------------------------------
+----+-------+---+------+----------+-------------------+
| NSS|    Nom|DID|  DNom|     Fecha|                _id|
+----+-------+---+------+----------+-------------------+
|4567| Marcos| 40|Gastro|01-09-2022|640cbabdf9972564d...|
|5678|  Sonia| 50|Gineco|01-09-2022|640cbac8f9972564d...|
|6789|Eduardo| 10|Cardio|01-09-2022|640cbad3f9972564d...|
+----+-------+---+------+----------+-------------------+


-------------------------------------------
Batch: 3
-------------------------------------------
+----+-----+---+------+----------+-------------------+
| NSS|  Nom|DID|  DNom|     Fecha|                _id|
+----+-----+---+------+----------+-------------------+
|1234|María| 10|Cardio|01-09-2022|640cbadcf9972564d...|
+----+-----+---+------+----------+-------------------+
```

As we have used the continuous trigger type with a 30 s interval, the data is not streamed as it is registered, but every 30 s; otherwise, you could not see aggregated data in different batches, unless you would be able to type quicker that the server is able to process the information.

Now, after we have seen several stream sources, it is time to deal with data storage. In data streaming terminology, those stores are known as data sinks.

8.2 Spark Streaming Data Sinks

As we have seen so far, Apache Spark Streaming is composed of three major logical components: a data source (input source), the processing engine (business logic), and finally an output destination (sink) for the resulting information after all the computations, aggregations, transformations, etc. have been performed.

Thus, in Spark Streaming, output sinks are used to save business logic applied to an external source. We have already seen in previous chapters that Spark Streaming uses the class `org.apache.spark.sql.streaming.DataStreamWriter` as an interface to write a streaming DataFrame/dataset to external storage systems via the `writeStream` method.

Spark includes a set of built-in output sinks:

- Console sink

- File sink

- Kafka sink

- ForeachBatch sink

- Foreach sink

All of them are natively supported by Spark Structured Streaming. The first one, the console sink, is mostly intended for testing and debugging as it does not support fault tolerance. The first three (console, file, and Kafka sinks) are already defined output formats: console, as in `format("console")`; file, as in `format("csv")`(or "json", "orc", or "parquet"); and "kafka". But what about writing the stream output to an arbitrary storage system like a NoSQL database like MongoDB or to a relational database like PostgreSQL? That is when Foreach and ForeachBatch sinks come into play.

Next, we study each one of them in detail.

Writing Streaming Data to the Console Sink

As you have already seen in our examples, the results are displayed onto the console. This data sink is not primarily intended for production systems; it would rather be a useful development and debugging tool.

Writing Streaming Data to the File Sink

The file sink stores the output data to a file system directory. Different file formats are supported such as JSON, ORC, CSV, and Parquet.

Here is an example of how you can save your streaming output to a CSV file. The code snippet is a modification of our streaming Hospital Queue Management System application to save processed data to a file system instead of outputting it to the console:

```
PatientDF.writeStream
    // You have to change this part of the code
    .format("csv")
    .option("path", "/tmp/streaming_output/csv")
    // … for this
    .format("parquet")
    .option("path", "/tmp/streaming_output/parquet")
    // ...
    .trigger(Trigger.ProcessingTime("5 seconds"))
    .option("checkpointLocation", checkpointDir)
    .outputMode("append")
    .option("truncate",false)
    .option("newRows",30)
    .start()
    .awaitTermination()
```

Now, if you have a look at the designated output directories, you should find an output similar to the one depicted in Figure 8-8.

```
/tmp/streaming_output/csv
├── part-00000-17b97ca1-0f4d-4e07-84fc-84ab6f499614-c000.csv
├── part-00000-4ffc3000-9f36-4527-9605-187c619e0f22-c000.csv
├── part-00000-ba9f1403-07aa-455b-b509-192301a962a9-c000.csv
├── part-00001-24059e92-f0c8-49a9-b84d-1b68b3e4ac6b-c000.csv
├── part-00001-292f6578-7d20-4fb2-987f-a43ceb79a66c-c000.csv
├── part-00002-9c5d1dbe-0f79-4956-93a5-a4b009b50267-c000.csv
├── part-00002-d6785701-8d2f-44a0-8554-94cca6ba6bd6-c000.csv
├── part-00003-3cab9a5f-bb17-4a39-a43f-3b1d5cbb9d83-c000.csv
├── part-00003-b1829cec-b66c-4058-a89a-bcb280b6b314-c000.csv
├── part-00004-248fb769-eb41-4d95-a28a-da5e299a94f2-c000.csv
├── part-00004-a1b3114b-707b-4c8d-9305-1e4365f8ccad-c000.csv
├── part-00005-677a90c5-e5c9-4260-9dbf-b76fe8dbd180-c000.csv
├── part-00005-c6c5f065-39a8-4475-bf95-46fa4225e256-c000.csv
├── part-00006-322ebef7-c2f3-423d-b36a-f8f709911ac1-c000.csv
├── part-00006-63621a58-68a5-413b-bc8a-90b1319c84bb-c000.csv
├── part-00007-2d7de7bc-970b-4160-9de5-44cbd6e595e9-c000.csv
├── part-00007-9a58865b-2e0f-42a2-88c1-582b02038617-c000.csv
├── part-00008-0777a663-445e-4409-886f-181180bd57e2-c000.csv
├── part-00008-f466fd68-1ea9-4add-853d-37eeab005b16-c000.csv
├── part-00009-9fe82a8f-fd51-4c80-ae22-357e3c1322ed-c000.csv
├── part-00009-edf2deb0-5fc0-4e72-8358-222602bf660c-c000.csv
├── part-00010-4a4725a3-9c2e-4da4-b429-a59b71fddc52-c000.csv
├── part-00010-f523215b-8d21-4f1d-bce6-776b6c24be98-c000.csv
├── part-00011-6b8e1a1f-01a4-449a-9972-ebd4ecd73faf-c000.csv
├── part-00011-ce91baac-bb88-44ec-8ee9-34e63a32c752-c000.csv
├── part-00012-3f309755-6c32-45bb-938b-bccf381aecd3-c000.csv
├── part-00012-58ceca22-8a63-4186-8abc-e34468287d2a-c000.csv
├── part-00013-7d89c945-e11e-4597-9c62-0b51cb77d214-c000.csv
├── part-00013-9ff93650-7824-4a11-8045-780d19656ae8-c000.csv
├── part-00014-6304077d-c156-4aa8-93ca-e5f9b9525729-c000.csv
├── part-00014-7d9d9bf3-1978-45b6-ac78-1436f3f080da-c000.csv
└── _spark_metadata
    ├── 4
    ├── 5
    └── 6
```

```
/tmp/streaming_output/parquet
├── part-00000-4c8d54e9-ed69-4c4c-90cc-d3108e863ff2-c000.snappy.parquet
├── part-00000-5bc7932f-d3b2-4ebb-8bc3-79c8f0cde817-c000.snappy.parquet
├── part-00001-1c035bdc-78be-4cfa-a395-efc7a38e009c-c000.snappy.parquet
├── part-00002-a81dd95b-351c-4a3f-abf6-cad7e861978d-c000.snappy.parquet
├── part-00003-44ab0689-22e4-4dc4-8b8c-768d8ba477e8-c000.snappy.parquet
├── part-00004-0eaa09eb-bd1f-4397-9790-c14cab24fb44-c000.snappy.parquet
├── part-00005-8fc01065-2668-4abd-a7b8-aee3ca2501fd-c000.snappy.parquet
├── part-00006-c5d1593b-df29-4cbd-b7f0-ff08e80c80f1-c000.snappy.parquet
├── part-00007-e4fc62ca-58d7-44e2-ba78-2e9cf423110f-c000.snappy.parquet
├── part-00008-38646ba1-979a-47f4-8c7a-1b6d41d534b9-c000.snappy.parquet
├── part-00009-745557ee-5cbc-42bd-bd1e-02e5bbf60e4f-c000.snappy.parquet
├── part-00010-1df235ee-e6b6-44b8-9ed2-15978236332f-c000.snappy.parquet
├── part-00011-d1fe3033-8b0f-4b11-baad-24fd0e284eae-c000.snappy.parquet
├── part-00012-dbca87c1-0d9c-4b5e-bac6-c2540ec6c91c-c000.snappy.parquet
├── part-00013-279b1d32-1ed3-4622-9149-829a6436af5c-c000.snappy.parquet
├── part-00014-45f85415-b850-4aea-a4aa-b1bc5b01c918-c000.snappy.parquet
├── part-00015-15235d5f-6a3c-4707-9313-3ad5b6fe79de-c000.snappy.parquet
├── part-00016-6390a0eb-c62e-4dbe-be47-57adde3f9aa3-c000.snappy.parquet
├── part-00017-9a0f2819-9e94-4dd3-af2c-9dafe0e7356d-c000.snappy.parquet
├── part-00018-a0fe3458-c09e-4d7e-b31e-7fb40d49c829-c000.snappy.parquet
├── part-00019-fe087c19-9649-4ce3-8375-8ec2337ee76a-c000.snappy.parquet
└── _spark_metadata
    ├── 7
    └── 8
```

Figure 8-8. *Example of streaming output to the file sink in CSV and Parquet formats*

For the sake of simplicity and visibility, in Figure 8-8 we have paired both outputs together. The CSV output format is on the left, and the Parquet output format is on the right.

Writing Streaming Data to the Kafka Sink

The Kafka sink publishes the output to one or more topics in Kafka.

Here is an example of how you can save your streaming output to a Kafka topic or topics:

```
counts.writeStream
 .format("kafka")
 .option("kafka.bootstrap.servers","host1:port1,host2:port2")
 // ...
 .option("topic", "patient")
 .option("checkpointLocation", "/tmp/kafka_checkpoint")
 .start()
 .awaitTermination()
```

Pay attention to the .option("kafka.bootstrap.servers","host1:port1,host2 :port2") line. As you can see, bootstrap.servers is a comma-separated list of socket connections (host and port pairs) corresponding to the IP addresses of the Kafka brokers in a "bootstrap" Kafka cluster. They are used by Kafka clients (producers and consumers) to connect to Kafka clusters.

So far we took advantage of sinks where the output format was already natively (built-in) supported like file, Kafka, or console. Now, we are going to study how to apply our own business logic to each stream record before saving it and how to write the information to our own defined data store using foreachBatch and foreach data sinks.

The main difference between the both of them is that while foreachBatch performs custom logic at the micro-batch level, foreach performs that custom logic at the row level.

Let's now study those two sinks.

Writing Streaming Data to the ForeachBatch Sink

The ForeachBatch sink takes a user-defined function that is executed on the output data for every micro-batch of a streaming query, for example:

```
def saveToCSV = (df: DataFrame, timeStamp: Long) => {
    df.withColumn("timeStamp", date_format(current_date(),"yyyyMMdd"))
    .write.format("csv")
    .option("path", "/tmp/streaming_output/foreachBatch")
    .mode("append")
    .save()
}
// ...
// ...
PatientDF.writeStream
    .trigger(Trigger.ProcessingTime("5 seconds"))
    .option("checkpointLocation", checkpointDir)
    .outputMode("append")
    .foreachBatch(saveToCSV)
    .start()
    .awaitTermination()
```

As you can see, this code snippet is a small modification of our previous examples. First of all, we have defined our own writing business logic encapsulated inside the saveToCSV() function. This function adds a timestamp to each micro-batch processed.

Here is the code example:

```
// File Sink to CSV

import org.apache.spark.sql.SparkSession
import org.apache.spark.sql.functions._
import org.apache.spark.sql.types.{IntegerType, StringType, StructField,
StructType,DoubleType,LongType}
import org.apache.spark.sql.{DataFrame, Dataset, Encoders, SparkSession}
import java.io.IOException
import org.apache.spark.sql.streaming._
import org.apache.spark.sql.streaming.{GroupState,GroupStateTimeout,
OutputMode}
import org.apache.spark.sql.DataFrame

val PatientsSchema = StructType(Array(
      StructField("NSS", StringType),
      StructField("Nom", StringType),
      StructField("DID", IntegerType),
      StructField("DNom", StringType),
      StructField("Fecha", StringType))
      )

case class Patient(
      NSS: String,
      Nom: String,
      DID: Option[Long],
      DNom: String,
      Fecha: String
)

def saveToCSV = (df: DataFrame, timeStamp: Long) => {
      df.withColumn("timeStamp", date_format(current_date(),"yyyyMMdd"))
      .write.format("csv")
      .option("path", "/tmp/streaming_output/foreachBatch")
```

```scala
      .mode("append")
      .save()
}

val spark:SparkSession = SparkSession.builder()
      .master("local[*]")
      .appName("Hand-On-Spark3_Socket_Data_Source")
      .getOrCreate()

spark.sparkContext.setLogLevel("ERROR")

import spark.implicits._

val host = "localhost"
val port = 9999
val checkpointDir = "/tmp/streaming_checkpoint"

try {
      val PatientDS = spark.readStream
      .format("socket")
      .option("host",host)
      .option("port",port)
      .load()
      .select(from_json(col("value"), PatientsSchema).as("patient"))
      .selectExpr("Patient.*")
      .as[Patient]

      printf("\n Listening and ready... \n")

      val PatientDF = PatientDS.select("*")

      PatientDF.writeStream
      .trigger(Trigger.ProcessingTime("5 seconds"))
      .option("checkpointLocation", checkpointDir)
      .outputMode("append")
      .foreachBatch(saveToCSV)
      .start()
      .awaitTermination()

} catch {
```

```
      case e: java.net.ConnectException => println("Error establishing
      connection to " + host + ":" + port)
      case e: IOException => println("IOException occurred")
      case t: Throwable => println("Error receiving data", t)
}finally {
      println("In finally block")
}
```

Now, open a terminal session and as usual type

```
nc -lk 9999
```

Then, run the preceding code example, and when you see the following in your notebook

```
Listening and ready...
```

go back to the previous terminal session and paste the JSON examples we provided you in Chapter 6, for instance:

```
{"NSS":"1234","Nom":"María", "DID":10, "DNom":"Cardio",
"Fecha":"01-09-2022"}
. . .
{"NSS":"2345","Nom":"Emilio", "DID":20, "DNom":"Neuro",
"Fecha":"01-09-2022"}
{"NSS":"3456","Nom":"Marta", "DID":30, "DNom":"Endo", "Fecha":"01-09-2022"}
```

After running the previous program and pasting the data to the terminal console, if you have a look at the designated output directory path /tmp/streaming_output/ foreachBatch/, you should find a bunch of files similar to the following:

/tmp/streaming_output/foreachBatch/
```
├── part-00000-07c12f65-b1d6-4c7b-b50d-2d8b25d724b8-c000.csv
├── part-00000-63507ff8-a09a-4c8e-a526-28890c170d96-c000.csv
├── part-00000-9a2caabe-7d84-4799-b788-a633cfc32042-c000.csv
├── part-00000-dabb8320-0c0e-4bb5-ad19-c36a53ac8d1e-c000.csv
├── part-00000-df0c4ba0-a9f0-40ed-b773-b879488b0a85-c000.csv
├── part-00000-f924d5cc-8e4a-4d5f-91b7-965ce2ac8710-c000.csv
├── part-00000-fd07c2e4-1db1-441c-8199-a69a064efe75-c000.csv
```

```
├──  part-00001-07c12f65-b1d6-4c7b-b50d-2d8b25d724b8-c000.csv
├──  part-00001-63507ff8-a09a-4c8e-a526-28890c170d96-c000.csv
├──  part-00001-9a2caabe-7d84-4799-b788-a633cfc32042-c000.csv
├──  part-00001-df0c4ba0-a9f0-40ed-b773-b879488b0a85-c000.csv
├──  part-00001-fd07c2e4-1db1-441c-8199-a69a064efe75-c000.csv
├──  part-00002-07c12f65-b1d6-4c7b-b50d-2d8b25d724b8-c000.csv
├──  part-00002-9a2caabe-7d84-4799-b788-a633cfc32042-c000.csv
├──  part-00002-df0c4ba0-a9f0-40ed-b773-b879488b0a85-c000.csv
├──  part-00002-fd07c2e4-1db1-441c-8199-a69a064efe75-c000.csv
├──  part-00003-07c12f65-b1d6-4c7b-b50d-2d8b25d724b8-c000.csv
├──  part-00003-df0c4ba0-a9f0-40ed-b773-b879488b0a85-c000.csv
├──  part-00003-fd07c2e4-1db1-441c-8199-a69a064efe75-c000.csv
├──  part-00004-07c12f65-b1d6-4c7b-b50d-2d8b25d724b8-c000.csv
├──  part-00004-df0c4ba0-a9f0-40ed-b773-b879488b0a85-c000.csv
├──  part-00005-07c12f65-b1d6-4c7b-b50d-2d8b25d724b8-c000.csv
├──  part-00005-df0c4ba0-a9f0-40ed-b773-b879488b0a85-c000.csv
└──  _SUCCESS
```

Remember, Spark by default writes to disk in a distributed manner; therefore, you are going to find the general program output as a sequence of partitioned files.

For example, we have copied and pasted the following JSON string into our console session:

```
{"NSS":"1009","Nom":"Julia", "DID":20, "DNom":"Neuro",
"Fecha":"01-09-2022"}
```

Now, if we open the file part-00000-07c12f65-b1d6-4c7b-b50d-2d8b25d724b8-c000.csv, for example

```
vi part-00000-07c12f65-b1d6-4c7b-b50d-2d8b25d724b8-c000.csv
```

we see the following content, including the timestamp at the end of the record, as we expected:

```
1009,Julia,20,Neuro,01-09-2022,20230317
```

Exactly the same could be seen opening other output files:

```
{"NSS":"2345","Nom":"Emilio", "DID":20, "DNom":"Neuro",
"Fecha":"01-09-2022"}

2345,Emilio,20,Neuro,01-09-2022,20230317

{"NSS":"4567","Nom":"Marcos", "DID":40, "DNom":"Gastro",
"Fecha":"01-09-2022"}

4567,Marcos,40,Gastro,01-09-2022,20230317
```

And so forth.

In a similar way, you could write your own function to use PostgreSQL as a data sink. You code could look like this:

```
def savePostgreSql = (df: DataFrame, timeStamp: Long) => {
    val url = "jdbc:postgresql://<host>:5432/database"

    df
    .withColumn("timeStamp", date_format(current_date(),"yyyyMMdd"))
    .write.format("jdbc")
    .option("driver": "org.postgresql.Driver")
    .option("url", url)
    .option("dbtable", "<your_table>")
    .option("user", "<your_user>")
    .option("password", <your_pasword>)
    .mode("append")
    .save()
}
```

Summarizing, foreachBatch writes each micro-batch to our designated storage applying our custom logic.

Writing Streaming Data to the Foreach Sink

The Foreach sink permits the application of user-defined business logic on each row during the data writing process. It can be used to write stream data to any kind of storage. If for any reason we cannot use foreachBatch, because a specific batch data writer does not exist or you need to use the continuous processing mode, then `foreach` could be the solution. When using `foreach` we have to implement three methods (`open`, `process`, and `close`):

- `open`: Is the function in charge of opening the connection

- `process`: Writes data to the designated connection

- `close`: Is the function responsible for closing the connection

Spark Structured Streaming implements the preceding methods in the following sequence: Method `open()` is called for every partition (`partition_id`) for every streaming batch/epoch (`epoch_id`) as `open(partitionId, epochId)`.

If `open(partitionId, epochId)` returns true for every row in the partition and for every batch/epoch, then the method `process(row)` is executed.

The method `close(error)` is called if any error appears while processing the data rows.

On the other hand, the `close()` method is executed if any `open()` method exists and returns successfully, provided no system failure occurred in between[1]:

```
counts.writeStream
.foreach( "some user logic goes here")
// ...
.start()
.awaitTermination()
```

Let's see now with a simple example how `foreach` can be implemented. For the purpose of this example, we have slightly modified our previous code snippet used for the `foreachBatch` sink to accommodate it to meet our necessities:

```
// Console Sink with foreach()

import org.apache.spark.sql.{Column, Row, SparkSession}
import org.apache.spark.sql.functions._
import org.apache.spark.sql.types.{IntegerType, StringType, StructField,
StructType,DoubleType,LongType}
import org.apache.spark.sql.{DataFrame, Dataset, Encoders, SparkSession}
```

---

[1] More information can be found here: `https://docs.databricks.com/structured-streaming/foreach.html`

```
import java.io.IOException
import org.apache.spark.sql.streaming._
import org.apache.spark.sql.streaming.{GroupState,GroupStateTimeout,
OutputMode}
import org.apache.spark.sql.{DataFrame,ForeachWriter}

val PatientsSchema = StructType(Array(
     StructField("NSS", StringType),
     StructField("Nom", StringType),
     StructField("DID", IntegerType),
     StructField("DNom", StringType),
     StructField("Fecha", StringType))
     )

case class Patient(
     NSS: String,
     Nom: String,
     DID: Option[Long],
     DNom: String,
     Fecha: String
)

val customWriterToConsole - new ForeachWriter[Row] {

     override def open(partitionId: Long, version: Long) = true

     override def process(record: Row) = {
     // You can transform record into a Sequence a loop through it
     //record.toSeq.foreach{col => println(col) }

     // ... or you can just print record field by field
     println("NSS: " + record.getAs("NSS")
          +" Nom: "  + record.getAs("Nom")
          +" DID: "  + record.getAs("DID")
          +" DNom: "  + record.getAs("DNom")
          +" Fecha : "  + record.getAs("Fecha"))
     }
```

```scala
      override def close(errorOrNull: Throwable) = {}
}

val spark:SparkSession = SparkSession.builder()
      .master("local[*]")
      .appName("Hand-On-Spark3_Socket_Data_Source")
      .getOrCreate()

spark.sparkContext.setLogLevel("ERROR")

import spark.implicits._

val host = "localhost"
val port = 9999
val checkpointDir = "/tmp/streaming_checkpoint"

try {
      val PatientDS = spark.readStream
      .format("socket")
      .option("host",host)
      .option("port",port)
      .load()
      .select(from_json(col("value"), PatientsSchema).as("patient"))
      .selectExpr("Patient.*")
      .as[Patient]

      printf("\n Listening and ready... \n")

      val PatientDF = PatientDS.select("*")

      PatientDF.writeStream
      .trigger(Trigger.ProcessingTime("5 seconds"))
      .option("checkpointLocation", checkpointDir)
      .outputMode("append")
      .foreach(customWriterToConsole)
      .start()
      .awaitTermination()
```

```
} catch {
    case e: java.net.ConnectException => println("Error establishing
    connection to " + host + ":" + port)
    case e: IOException => println("IOException occurred")
    case t: Throwable => println("Error receiving data", t)
}finally {
    println("In finally block")
}
```

Before executing the preceding code, open a terminal session and create a socket session as follows:

```
$ nc -lk 9999
```

Once the socket session has been created, it is time to run the code. As soon as you see the line Listening and ready... on your screen, go back to the terminal with the socket session open and start typing JSON lines. You can use lines like following:

```
{"NSS":"1234","Nom":"María", "DID":10, "DNom":"Cardio", "Fecha":"01-09-2022"}
{"NSS":"2345","Nom":"Emilio", "DID":20, "DNom":"Neuro", "Fecha":"01-09-2022"}
{"NSS":"3456","Nom":"Marta", "DID":30, "DNom":"Endo", "Fecha":"01-09-2022"}
{"NSS":"4567","Nom":"Marcos", "DID":40, "DNom":"Gastro", "Fecha":"01-09-2022"}
{"NSS":"5678","Nom":"Sonia", "DID":50, "DNom":"Gineco", "Fecha":"01-09-2022"}
{"NSS":"6789","Nom":"Eduardo", "DID":10, "DNom":"Cardio", "Fecha":"01-09-2022"}
{"NSS":"1001","Nom":"Lorena", "DID":10, "DNom":"Cardio", "Fecha":"01-09-2022"}
{"NSS":"1006","Nom":"Sara", "DID":20, "DNom":"Neuro", "Fecha":"01-09-2022"}
{"NSS":"1002","Nom":"Teresa", "DID":10, "DNom":"Cardio", "Fecha":"01-09-2022"}
{"NSS":"1003","Nom":"Luis", "DID":20, "DNom":"Neuro", "Fecha":"01-09-2022"}
```

You will see an output like this coming out of you program:

```
 Listening and ready...
NSS: 1234 Nom: María DID: 10 DNom: Cardio Fecha : 01-09-2022
NSS: 2345 Nom: Emilio DID: 20 DNom: Neuro Fecha : 01-09-2022
NSS: 3456 Nom: Marta DID: 30 DNom: Endo Fecha : 01-09-2022
NSS: 4567 Nom: Marcos DID: 40 DNom: Gastro Fecha : 01-09-2022
NSS: 5678 Nom: Sonia DID: 50 DNom: Gineco Fecha : 01-09-2022
NSS: 6789 Nom: Eduardo DID: 10 DNom: Cardio Fecha : 01-09-2022
```

```
NSS: 1001 Nom: Lorena DID: 10 DNom: Cardio Fecha : 01-09-2022
NSS: 1006 Nom: Sara DID: 20 DNom: Neuro Fecha : 01-09-2022
NSS: 1002 Nom: Teresa DID: 10 DNom: Cardio Fecha : 01-09-2022
NSS: 1003 Nom: Luis DID: 20 DNom: Neuro Fecha : 01-09-2022
```

Going back to the previous code example, you can see that the only differences
are the customWriterToConsole() function implementing the ForeachWriter and the
foreach sink call itself, inside the writeStream method.

Notice the implementation of the three mandatory methods—open, process, and
finally close:

```
val customWriterToConsole = new ForeachWriter[Row] {

    override def open(partitionId: Long, version: Long) = true

    override def process(record: Row) = {
    // You can transform record into a Sequence a loop through it
    //record.toSeq.foreach{col => println(col) }

    // ... or you can just print record field by field
    println("NSS: " + record.getAs("NSS")
           +" Nom: "   + record.getAs("Nom")
           +" DID: "   + record.getAs("DID")
           +" DNom: "   + record.getAs("DNom")
           +" Fecha : "   + record.getAs("Fecha"))
    }

    override def close(errorOrNull: Throwable) = {}
}
```

And notice the foreach sink call inside the writeStream method:

```
PatientDF.writeStream
    .trigger(Trigger.ProcessingTime("5 seconds"))
    .option("checkpointLocation", checkpointDir)
    .outputMode("append")
    .foreach(customWriterToConsole)
    .start()
    .awaitTermination()
```

You can modify the `customWriterToConsole` implementation to meet your particular needs.

Writing Streaming Data to Other Data Sinks

In previous sections we have seen how to use the Spark Structured Streaming built-in data sinks. In this section we are going to see how to use the MongoDB Spark connector to stream live data to MongoDB.

MongoDB stores the information in JSON-like documents with a variable structure, offering a dynamic and flexible schema. MongoDB was designed for high availability and scalability and natively incorporates built-in replication and auto-sharding.

The MongoDB sink allows you to write events from Spark to a MongoDB instance. The sink connector converts the Spark streaming event data into a MongoDB document and will do an append or `overwrite` depending on the save mode configuration you choose.

The MongoDB sink connector expects the output database created up front, while the destination MongoDB collections can be created at runtime if they do not exist.

A graphical representation of the MongoDB connector for Spark can be seen in Figure 8-9.

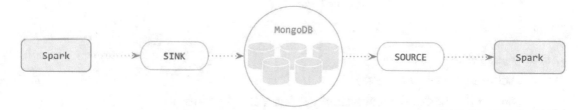

Figure 8-9. *MongoDB connector for Spark representation*

Next, we are going to see a practical code example, showing how to use MongoDB as a Spark data sink. As in other previous examples, the program reads JSON files from a directory as soon as they emerge over there and inserts the data into a MongoDB collection. The JSON files we are using in this example are the *patient* examples we have been using so far in previous examples and the ones firstly included in Chapter 6:

```
import org.apache.spark.sql.SparkSession
import org.apache.spark.{SparkConf, SparkContext}
import org.apache.spark.sql.types.{IntegerType, StringType, StructField,
StructType}
```

```scala
val spark:SparkSession = SparkSession
    .builder()
    .config("spark.jars.packages", "org.mongodb.spark:mongo-spark-
    connector:10.1.1")
    .master("local[*]")
    .appName("Hand-On-Spark3_File_Data_Source_MongoDB_Sink")
    .getOrCreate()

val sc = spark.sparkContext

sc.setLogLevel("ERROR")

val mongoDBURI = "mongodb+srv://<user>:<password>@hands-on-spark3.akxgvpe.
mongodb.net/?retryWrites=true&w=majority"

val PatientsSchema = StructType(Array(
    StructField("NSS", StringType),
    StructField("Nom", StringType),
    StructField("DID", IntegerType),
    StructField("DNom", StringType),
    StructField("Fecha", StringType))
    )

val df = spark.readStream
    .schema(PatientsSchema)
    .option("checkpointLocation", "/tmp/checkpoint")
    .json("/tmp/stream_mongo")

df.printSchema()

val newDF = df.select("*") // Here you could transform your data

newDF.printSchema()

newDF.writeStream
    .format("mongodb")
    .option("checkpointLocation", "/tmp/checkpoint")
    .option("forceDeleteTempCheckpointLocation", "true")
```

To see the code working, create a streaming directory, /tmp/stream_mongo, for example, where to copy your JSON files.

When you run the preceding code and you see the following output

```
root
|-- NSS: string (nullable = true)
|-- Nom: string (nullable = true)
|-- DID: integer (nullable = true)
|-- DNom: string (nullable = true)
|-- Fecha: string (nullable = true)
```

you can start copying files to the designated streaming directory. For the purpose of this example, we use the JSON files we used in Chapter 6. Here is an example of how you can do it:

```
$ cp /tmp/json/patient1.json /tmp/stream_mongo
$ cp /tmp/json/patient2.json /tmp/stream_mongo
$ cp /tmp/json/patient3.json /tmp/stream_mongo
$ cp /tmp/json/patient4.json /tmp/stream_mongo
$ cp /tmp/json/patient5.json /tmp/stream_mongo
$ cp /tmp/json/patient6.json /tmp/stream_mongo
```

Remember the information inside those files looks like this:

```
{"NSS":"1009","Nom":"Julia", "DID":20, "DNom":"Neuro",
"Fecha":"01-09-2022"}
{"NSS":"1010","Nom":"Javier", "DID":30, "DNom":"Endo",
"Fecha":"01-09-2022"}
{"NSS":"1011","Nom":"Laura", "DID":50, "DNom":"Gineco",
"Fecha":"01-09-2022"}
{"NSS":"1012","Nom":"Nuria", "DID":10, "DNom":"Cardio",
"Fecha":"01-09-2022"}
{"NSS":"1013","Nom":"Helena", "DID":10, "DNom":"Cardio",
"Fecha":"01-09-2022"}
{"NSS":"1014","Nom":"Nati", "DID":10, "DNom":"Cardio",
"Fecha":"01-09-2022"}
```

Now, if you have a look at your MongoDB database—in our case we have used the graphical interface MongoDB Compass to do it—you could see the data inserted from the streaming process.

Figure 8-10 shows you how to filter the already recorded data using different data keys. In this case we have used the department ID ("DID"). Remember MongoDB stores the information in a JSON-like format, not in tables as traditional OLTP databases do.

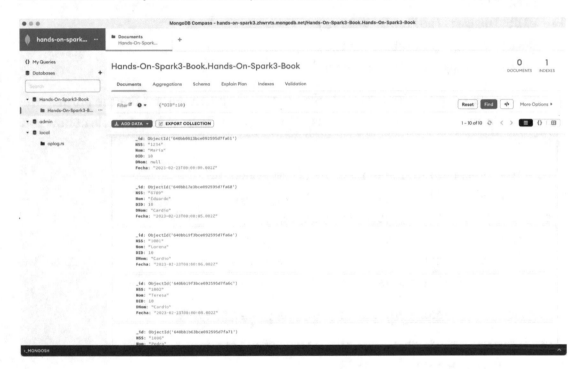

Figure 8-10. *MongoDB Compass filtering data by department ID (DID)*

In Figure 8-11 you can see a similar filtering query, but in this case we have filtered by the Social Security Number (SSN).

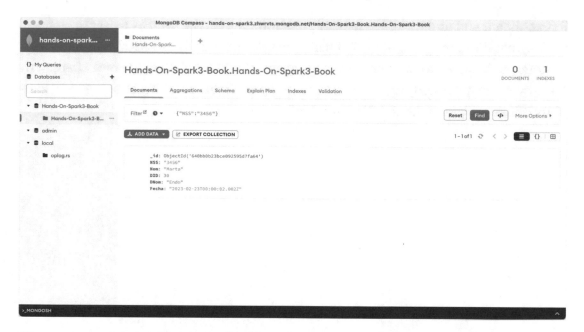

Figure 8-11. *MongoDB Compass filtering data by Social Security Number (SSN)*

Wrapping up, MongoDB is a well-known document-oriented, nonrelational database intended for use with semi-structured data. It is very flexible and can handle large volumes of heterogeneous information. Both MongoDB and Spark are published under a free and open source license and together constitute a solid pillar to consider in any modern data architecture.

8.3 Summary

In this chapter we went over the Spark Structured Streaming module. In particular we have studied the most common data sources and data sinks, regarding streaming data processing. Firstly, we studied the built-in Spark Structured Streaming data sources, paying special attention to the most typical ones: the file, socket, and Kafka sources. Kafka is one of most important streaming frameworks nowadays; therefore, we developed a specific code example showing how to use it as a live stream source. Secondly, we showed how to implement a custom data source and implemented another practical example how to do it with MongoDB. After that, we moved forward and repeated the same process with data sinks. First, we went through the defined data sinks, that is to say, the console sink, file sink, and Kafka sink. Later on, we studied the foreachBatch and foreach sinks and analyzed how they can be used by a user to create their own tailor-made data sinks. To finalize, we also provided a practical example of a custom-made data sink implemented once again with MongoDB. In the next chapter, we are moving forward studying advanced streaming configurations, introducing the Event-Time Window Operations and Watermarking.

CHAPTER 9

Event-Time Window Operations and Watermarking

After having studied the insights of Apache Spark Streaming and Structured Streaming, in this chapter, we are going to focus on time-based stream processing.

Data analytics is evolving from batch to stream data processing for many use cases. One of the reasons for this shift is that it is becoming more and more commonly accepted that streaming data is more suited to model the life we live. This is particularly true when we think about most of the systems we want to analyze and model—autonomous cars receiving and emitting satellite navigation coordinates, Internet of things (IoT) devices exchanging signals, road sensors counting vehicles for traffic control, wearable devices, etc.—all have a common similarity; they all appear as a continuous stream of events and in a timely manner. In fact, streaming data sources are almost omnipresent.

Additionally, events are generated as a result of some activity, and in many scenarios they require some immediate action to be taken. Consider, for example, applications for fraud or anomaly detection or personalization, marketing, and advertising in real time as some of the most common use cases of real-time stream processing and event-driven applications.

Coherent time semantics are of paramount importance in stream processing as many operations in event processing such as aggregation over a time window, joins, and stragglers management depend on time.

In this chapter, we are going to go through the concept of temporal windows, also known as time windows, for stream processing, study Spark's built-in window functions, and explain windowing semantics.

© Alfonso Antolínez García 2023
A. Antolínez García, *Hands-on Guide to Apache Spark 3*, https://doi.org/10.1007/978-1-4842-9380-5_9

9.1 Event-Time Processing

As mentioned just before, many operations in real-time event stream processing are depending on time. When dealing with events and time, we have several options of time marks for event, and depending on the use case at hand, we must prioritize one variant over the others:

- Event-time: It refers to the time in which the event was created, for example, produced by a sensor.

- Ingestion-time: It denotes the moment in time when the event was ingested by the event streaming platform. It is implemented by adding a timestamp to the event when it enters the streaming platform.

- Processing-time, also called Wall-clock-time: It is the moment when the event is effectively processed.

Next, Figure 9-1 graphically explains the previous event-time processing concepts.

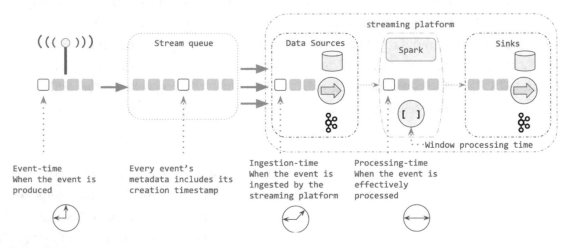

Figure 9-1. *Stream event-time processing schema*

9.2 Stream Temporal Windows in Apache Spark

In real-time stream processing, performing actions on the data contained in temporal windows is one of the most common operations.

Temporal windows, also known as time windows, group stream elements by time intervals. Apache Spark Structured Streaming also has event-time support and allows windowed computations over event time and native support for windowing functions. Before Apache version 3.2, Spark only supported "tumbling windows" and "sliding windows," but starting with Spark 3.2, Spark also includes "session windows" which can also be used for both streaming and batch queries enabling engineers to develop complex stream processing jobs with minimal work.

With Structured Streaming, data aggregations are very similar to Spark grouped aggregations when applied to sliding windows. Regarding grouped aggregations, aggregated calculations are maintained for each different element of the grouping column. When it comes to window-based aggregations, aggregated calculations are maintained for each window the event-time value belongs to.

Therefore, at the moment this book was written, Spark offers three types of temporal windows to choose from:

- Tumbling windows

- Sliding windows

- Session windows

The common denominator of the precedent window types is that they are applied over continuous streaming of data, splitting it into finite collections of information. The application of temporal or finite time windows to stream data is particularly indicated when we would like to perform operations like aggregations, joins, and pattern identification. The next sections describe how the tumbling, sliding, and session window types work and how to practically implement them. Let's study each one of them in detail.

What Are Temporal Windows and Why Are They Important in Streaming

Consider our example of the Hospital Queue Management System we have been using so far. Consider as well that we have a counter device counting every 15 seconds the number of patients entering a hospital. The resulting stream of events could result like in Figure 9-2.

Figure 9-2. *Example of counting device delivering a stream of events*

If we would like to know the number of patients entering the hospital, we could add up the number of patients counted. However, the nature of a stream is that we face a scenario of unbound data. That is to say, the flow of counts is endless, and therefore we cannot produce a final total of the number of patients entering the facilities. One alternative could be computing partial sums, that is to say, adding up the counts received and updating the partial sum with the new values as they are collected. Acting like this, we collect a series of running totals updated with new counts as it is shown in Figure 9-3.

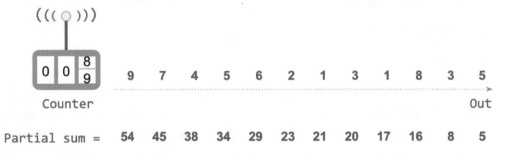

Figure 9-3. *Sequence of partial sums*

However, a sequence of partial sums is a live metric as it is constantly updated. Therefore, the strategy of rolling sums cannot be the best option if you want to analyze data variability over time, for example, when is there a bigger influx of patients to the hospital, in the morning or evening? Or how many patients enter the hospital every unit of time as we see in Figure 9-4.

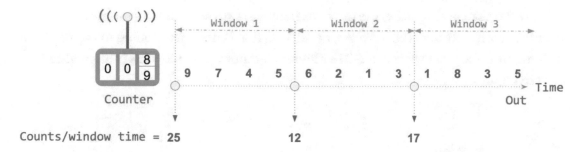

Figure 9-4. *Example of temporal window to count events per window time*

To answer questions like the previous ones, we have different kinds of temporary window operations. Next we are going to study the tumbling window operations.

9.3 Tumbling Windows

Tumbling windows or nonoverlapping windows discretize a stream into nonoverlapping segments of data and apply a function against them, like the example depicted in Figure 9-5.

The main features of tumbling windows are that disjuncts repeat, and an event only belongs to one, and only one, tumbling window.

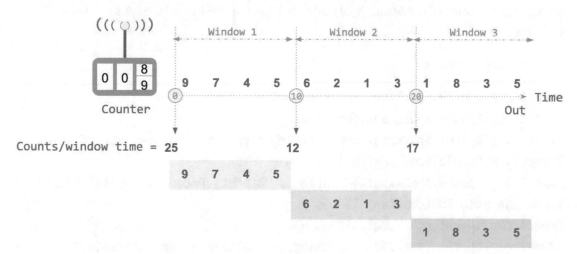

Figure 9-5. *A ten-second tumbling window*

For the sake of simplicity, the previous figures are shown the same number of events per window interval; however, be advised that is not always going to happen, and different numbers of events can fall in different temporary windows as is highlighted next in Figure 9-6.

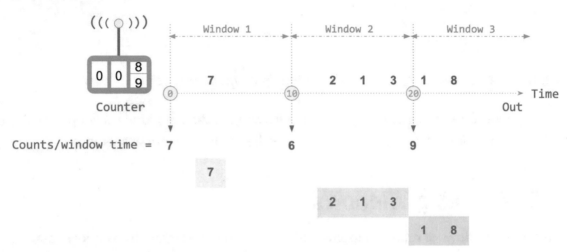

Figure 9-6. *A ten-second tumbling window with different number of events per window*

The next code snippet uses a tumbling window of ten seconds' size to perform an aggregate count of the number of patients entering the hospital over the same window time.

```
// Tumbling windows

import org.apache.spark.sql.SparkSession
import org.apache.spark.sql.functions._
import org.apache.spark.sql.types.{IntegerType, StringType, StructField,
StructType,DoubleType,LongType}
import org.apache.spark.sql.{DataFrame, Dataset, Encoders, SparkSession}
import java.io.IOException
import org.apache.spark.sql.streaming._
import org.apache.spark.sql.streaming.{GroupState,GroupStateTimeout,
OutputMode}
import org.apache.spark.sql.DataFrame
```

```scala
val PatientsSchema = StructType(Array(
    StructField("NSS", StringType),
    StructField("Nom", StringType),
    StructField("DID", IntegerType),
    StructField("DNom", StringType),
    StructField("Fecha", StringType))
    )

val spark:SparkSession = SparkSession.builder()
    .master("local[10]")
    .appName("Hand-On-Spark3_Socket_Data_Source")
    .getOrCreate()

spark.sparkContext.setLogLevel("ERROR")

import spark.implicits._

try {
    val PatientDS = spark.readStream
    .schema(PatientsSchema)
    .json("/tmp/window")

    printf("\n Listening and ready... \n")

    val PatientDF = PatientDS
    .groupBy(window(col("Fecha"), "10 seconds"))
    .agg(count("DNom").alias("Suma_x_Dpt"))

    PatientDF.writeStream
        .outputMode("complete")
        .format("console")
        .option("truncate", false)
        .start()
        .awaitTermination()
} catch {
    case e: IOException => println("IOException occurred")
    case t: Throwable => println("Error receiving data", t)
}finally {
    println("In finally block")
}
```

To run the previous code example, first of all you have to create the necessary data source (in our case "/tmp/window") to pull the corresponding JSON files to.

Ones you have done so, run the code and when you see the message

```
Listening and ready...
```

Start copying files to the data source, for example:

```
$ cp json_file1.json /tmp/window
$ cp json_file2.json /tmp/window
$ cp json_file3.json /tmp/window
$ cp json_file4.json /tmp/window
$ cp json_file5.json /tmp/window
$ cp json_file6.json /tmp/window
$ cp json_file7.json /tmp/window
```

You will have a similar output like the following one coming out of your program:

```
 Listening and ready...
------------------------------------------
Batch: 0
------------------------------------------
+-------------------------------------------+----------+
|window                                     |Suma_x_Dpt|
+-------------------------------------------+----------+
|{2023-02-23 01:00:00, 2023-02-23 01:00:10}|1         |
+-------------------------------------------+----------+

------------------------------------------
Batch: 1
------------------------------------------
+-------------------------------------------+----------+
|window                                     |Suma_x_Dpt|
+-------------------------------------------+----------+
|{2023-02-23 01:00:00, 2023-02-23 01:00:10}|3         |
+-------------------------------------------+----------+
```

```
------------------------------------------
Batch: 2
------------------------------------------
+------------------------------------------+----------+
|window                                    |Suma_x_Dpt|
+------------------------------------------+----------+
|{2023-02-23 01:00:00, 2023-02-23 01:00:10}|6         |
+------------------------------------------+----------+


------------------------------------------
Batch: 3
------------------------------------------
+------------------------------------------+----------+
|window                                    |Suma_x_Dpt|
+------------------------------------------+----------+
|{2023-02-23 01:00:00, 2023-02-23 01:00:10}|10        |
+------------------------------------------+----------+


------------------------------------------
Batch: 4
------------------------------------------
+------------------------------------------+----------+
|window                                    |Suma_x_Dpt|
+------------------------------------------+----------+
|{2023-02-23 01:00:00, 2023-02-23 01:00:10}|10        |
|{2023-02-23 01:00:10, 2023-02-23 01:00:20}|1         |
+------------------------------------------+----------+


------------------------------------------
Batch: 5
------------------------------------------
+------------------------------------------+----------+
|window                                    |Suma_x_Dpt|
+------------------------------------------+----------+
|{2023-02-23 01:00:00, 2023-02-23 01:00:10}|10        |
|{2023-02-23 01:00:10, 2023-02-23 01:00:20}|1         |
+------------------------------------------+----------+
```

```
-------------------------------------------
Batch: 6
-------------------------------------------
+-----------------------------------------+----------+
|window                                   |Suma_x_Dpt|
+-----------------------------------------+----------+
|{2023-02-23 01:00:00, 2023-02-23 01:00:10}|10       |
|{2023-02-23 01:00:10, 2023-02-23 01:00:20}|1        |
+-----------------------------------------+----------+
```

Now if you introduce a small change in the previous code like this

```
PatientDF.printSchema()
```

Before this part of the code

```
PatientDF.writeStream
.outputMode("complete")
.format("console")
.option("truncate", false)
.start()
.awaitTermination()
```

You will see the schema of your window data frame is like the following:

```
root
 |-- window: struct (nullable = true)
 |    |-- start: timestamp (nullable = true)
 |    |-- end: timestamp (nullable = true)
 |-- Suma_x_Dpt: long (nullable = false)
```

Therefore, if you prefer to see the window boundaries in separate columns, you can tweak the previous code as follows:

```
val PatientDF = PatientDS
.groupBy(window(col("Fecha"), "10 seconds"))
.agg(count("DNom").alias("Suma_x_Dpt"))
.select("window.start", "window.end", "Suma_x_Dpt")
```

And you will see the window information as shown in the following:

```
Listening and ready...
-------------------------------------------
Batch: 0
-------------------------------------------
+-------------------+-------------------+---------+
|start              |end                |Suma_x_Dpt|
+-------------------+-------------------+---------+
|2023-02-23 01:00:00|2023-02-23 01:00:10|1        |
+-------------------+-------------------+---------+

-------------------------------------------
Batch: 1
-------------------------------------------
+-------------------+-------------------+---------+
|start              |end                |Suma_x_Dpt|
+-------------------+-------------------+---------+
|2023-02-23 01:00:00|2023-02-23 01:00:10|3        |
+-------------------+-------------------+---------+

-------------------------------------------
Batch: 2
-------------------------------------------
+-------------------+-------------------+---------+
|start              |end                |Suma_x_Dpt|
+-------------------+-------------------+---------+
|2023-02-23 01:00:00|2023-02-23 01:00:10|6        |
+-------------------+-------------------+---------+

-------------------------------------------
Batch: 3
-------------------------------------------
+-------------------+-------------------+---------+
|start              |end                |Suma_x_Dpt|
+-------------------+-------------------+---------+
|2023-02-23 01:00:00|2023-02-23 01:00:10|10       |
+-------------------+-------------------+---------+
```

```
-----------------------------------------
Batch: 4
-----------------------------------------
+-----------------+-----------------+----------+
|start            |end              |Suma_x_Dpt|
+-----------------+-----------------+----------+
|2023-02-23 01:00:00|2023-02-23 01:00:10|10        |
|2023-02-23 01:00:10|2023-02-23 01:00:20|1         |
+-----------------+-----------------+----------+


-----------------------------------------
Batch: 5
-----------------------------------------
+-----------------+-----------------+----------+
|start            |end              |Suma_x_Dpt|
+-----------------+-----------------+----------+
|2023-02-23 01:00:00|2023-02-23 01:00:10|10        |
|2023-02-23 01:00:10|2023-02-23 01:00:20|1         |
+-----------------+-----------------+----------+


-----------------------------------------
Batch: 6
-----------------------------------------
+-----------------+-----------------+----------+
|start            |end              |Suma_x_Dpt|
+-----------------+-----------------+----------+
|2023-02-23 01:00:00|2023-02-23 01:00:10|10        |
|2023-02-23 01:00:10|2023-02-23 01:00:20|1         |
+-----------------+-----------------+----------+
```

9.4 Sliding Windows

In certain cases, we might require a different kind of window. For example, we may need overlapping windows if we would like to know every 30 minutes how many patients entered the hospital during the last minute. To answer this kind of question, we need the sliding windows.

Sliding windows like tumbling windows are "fixed-sized," but unlike them, they can overlap. When window overlapping happens, an event can belong to multiple windows. Overlapping occurs when the duration of the slide is smaller than the duration of the window.

Thus, in Spark Streaming, to define a sliding window, two parameters are needed: the window size (interval) and a sliding offset (overlapping dimension). For example, in Figure 9-7, we have created a sliding window with ten seconds of size and sliding offset of five seconds.

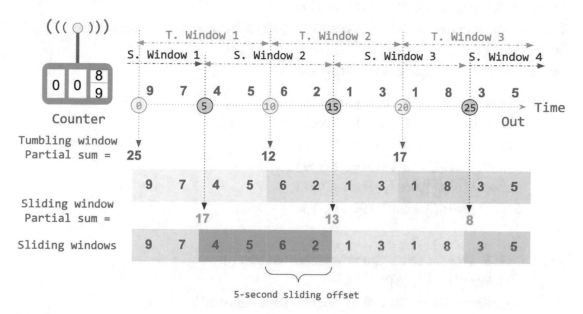

Figure 9-7. *A ten-second sliding windows with sliding offset of five seconds*

In the next code example, we calculate the aggregated number of people entering the hospital every ten seconds. The example illustrates how to create a sliding window on the column Fecha for every ten seconds and adding a sliding offset of five seconds.

Note   Please notice the time intervals established are very narrow for the sake of usage illustration. The same code applied to a real hospital will probably use wider time intervals.

```scala
// Sliding Windows

import org.apache.spark.sql.SparkSession
import org.apache.spark.sql.functions._
import org.apache.spark.sql.types.{IntegerType, StringType, StructField,
StructType,DoubleType,LongType}
import org.apache.spark.sql.{DataFrame, Dataset, Encoders, SparkSession}
import java.io.IOException
import org.apache.spark.sql.streaming._
import org.apache.spark.sql.streaming.{GroupState,GroupStateTimeout,
OutputMode}
import org.apache.spark.sql.DataFrame

val PatientsSchema = StructType(Array(
     StructField("NSS", StringType),
     StructField("Nom", StringType),
     StructField("DID", IntegerType),
     StructField("DNom", StringType),
     StructField("Fecha", StringType))
     )

val spark:SparkSession = SparkSession.builder()
     .master("local[10]")
     .appName("Hand-On-Spark3_Socket_Data_Source")
     .getOrCreate()

spark.sparkContext.setLogLevel("ERROR")

import spark.implicits._

try {
     val PatientDS = spark.readStream
     .schema(PatientsSchema)
     .json("/tmp/window")

     printf("\n Listening and ready... \n")

     val PatientDF = PatientDS
     .groupBy(window(col("Fecha"), "10 seconds", "5 seconds"))
```

```
      .agg(count("DID").alias("Suma_x_Dpt"))

      PatientDF.writeStream
      .outputMode("complete")
      .format("console")
      .option("truncate", false)
      .start()
      .awaitTermination()
} catch {
      case e: IOException => println("IOException occurred")
      case t: Throwable => println("Error receiving data", t)
}finally {
      println("In finally block")
}
```

As we did in our previous tumbling windows example, before trying to run the previous code example, first you have to create the data source for the JSON files ("/tmp/window" folder in our case). After that, you can start pouring JSON files to that directory. For example

```
$ cp json_file9.json /tmp/window
$ cp json_file8.json /tmp/window
$ cp json_file7.json /tmp/window
...
$ cp json_file1.json /tmp/window
```

As soon as you copy the mentioned files, and depending on the copying rate you apply, the program will create a window size of ten seconds with a sliding interval of five seconds. A new window of ten seconds will be created every 5, with a five-second gap from the beginning of the previous one, as it is shown in the next program output.

```
Listening and ready...
-------------------------------------------
Batch: 0
-------------------------------------------
```

```
+---------------------------------------+----------+
|window                                 |Suma_x_Dpt|
+---------------------------------------+----------+
|{2023-02-23 01:00:25, 2023-02-23 01:00:35}|5         |
|{2023-02-23 01:00:35, 2023-02-23 01:00:45}|4         |
|{2023-02-23 01:00:30, 2023-02-23 01:00:40}|9         |
+---------------------------------------+----------+


-------------------------------------------
Batch: 1
-------------------------------------------

+---------------------------------------+----------+
|window                                 |Suma_x_Dpt|
+---------------------------------------+----------+
|{2023-02-23 01:00:25, 2023-02-23 01:00:35}|10        |
|{2023-02-23 01:00:20, 2023-02-23 01:00:30}|8         |
|{2023-02-23 01:00:35, 2023-02-23 01:00:45}|4         |
|{2023-02-23 01:00:30, 2023-02-23 01:00:40}|9         |
|{2023-02-23 01:00:15, 2023-02-23 01:00:25}|3         |
+---------------------------------------+----------+


-------------------------------------------
Batch: 2
-------------------------------------------

+---------------------------------------+----------+
|window                                 |Suma_x_Dpt|
+---------------------------------------+----------+
|{2023-02-23 01:00:25, 2023-02-23 01:00:35}|10        |
|{2023-02-23 01:00:20, 2023-02-23 01:00:30}|10        |
|{2023-02-23 01:00:35, 2023-02-23 01:00:45}|4         |
|{2023-02-23 01:00:10, 2023-02-23 01:00:20}|5         |
|{2023-02-23 01:00:30, 2023-02-23 01:00:40}|9         |
|{2023-02-23 01:00:15, 2023-02-23 01:00:25}|10        |
+---------------------------------------+----------+
```

```
-----------------------------------------
Batch: 3
-----------------------------------------
+-----------------------------------------+----------+
|window                                   |Suma_x_Dpt|
+-----------------------------------------+----------+
|{2023-02-23 01:00:25, 2023-02-23 01:00:35}|10       |
|{2023-02-23 01:00:20, 2023-02-23 01:00:30}|11       |
|{2023-02-23 01:00:35, 2023-02-23 01:00:45}|4        |
|{2023-02-23 01:00:10, 2023-02-23 01:00:20}|10       |
|{2023-02-23 01:00:30, 2023-02-23 01:00:40}|9        |
|{2023-02-23 01:00:15, 2023-02-23 01:00:25}|16       |
+-----------------------------------------+----------+
```

As we did with tumbling windows, you can modify the previous code snippet as follows:

```
val PatientDF = PatientDS
.groupBy(window(col("Fecha"), "10 seconds", "5 seconds"))
.agg(count("DID").alias("Suma_x_Dpt"))
.select("window.start", "window.end", "Suma_x_Dpt")
```

to separate window time data in two different columns.

As in the tumbling windows example, you will get a similar output like the following one:

```
Listening and ready...
-----------------------------------------
Batch: 0
-----------------------------------------
+-------------------+-------------------+----------+
|start              |end                |Suma_x_Dpt|
+-------------------+-------------------+----------+
|2023-02-23 01:00:25|2023-02-23 01:00:35|5         |
|2023-02-23 01:00:35|2023-02-23 01:00:45|4         |
|2023-02-23 01:00:30|2023-02-23 01:00:40|9         |
+-------------------+-------------------+----------+
```

```
-----------------------------------------
Batch: 1
-----------------------------------------
+-----------------+-----------------+---------+
|start            |end              |Suma_x_Dpt|
+-----------------+-----------------+---------+
|2023-02-23 01:00:25|2023-02-23 01:00:35|10      |
|2023-02-23 01:00:20|2023-02-23 01:00:30|8       |
|2023-02-23 01:00:35|2023-02-23 01:00:45|4       |
|2023-02-23 01:00:30|2023-02-23 01:00:40|9       |
|2023-02-23 01:00:15|2023-02-23 01:00:25|3       |
+-----------------+-----------------+---------+


-----------------------------------------
Batch: 2
-----------------------------------------
+-----------------+-----------------+---------+
|start            |end              |Suma_x_Dpt|
+-----------------+-----------------+---------+
|2023-02-23 01:00:25|2023-02-23 01:00:35|10      |
|2023-02-23 01:00:20|2023-02-23 01:00:30|10      |
|2023-02-23 01:00:35|2023-02-23 01:00:45|4       |
|2023-02-23 01:00:10|2023-02-23 01:00:20|5       |
|2023-02-23 01:00:30|2023-02-23 01:00:40|9       |
|2023-02-23 01:00:15|2023-02-23 01:00:25|10      |
+-----------------+-----------------+---------+


-----------------------------------------
Batch: 3
-----------------------------------------
+-----------------+-----------------+---------+
|start            |end              |Suma_x_Dpt|
+-----------------+-----------------+---------+
|2023-02-23 01:00:25|2023-02-23 01:00:35|10      |
|2023-02-23 01:00:20|2023-02-23 01:00:30|11      |
|2023-02-23 01:00:35|2023-02-23 01:00:45|4       |
```

```
|2023-02-23 01:00:10|2023-02-23 01:00:20|10        |
|2023-02-23 01:00:30|2023-02-23 01:00:40|9         |
|2023-02-23 01:00:15|2023-02-23 01:00:25|16        |
+------------------+------------------+----------+
```

With sliding windows, we can answer questions such as what was the number of patients visiting our hospital during the last minute, hour, etc.? or trigger events such as "ring an alarm" whenever more than five patients for the same medical department enter the hospital in the last ten seconds.

In the next section, we are going to study session windows which have a different semantics compared to the previous two types of windows.

9.5 Session Windows

Session windows have an important different characteristic compared to tumbling and sliding windows. Session windows have a variable geometry. Session windows' length is dynamic in size depending on the incoming events.

Session windows gather events that arrive at similar moments in time, isolating periods of data inactivity. A session window starts with an input event collected and lasts for as long as we keep receiving data within the gap interval duration equivalent to the window length. Thus, in any case, it closes itself when the maximum window length is reached. For example, in our previous examples in which we had a window size of ten seconds, the session windows will begin right after receiving the first input. Thereafter, all the events acquired within ten seconds will be associated with that window. This window will close itself if it does not receive more inputs for a period of ten seconds. A graphical depiction of how a session window works can be seen in Figure 9-8.

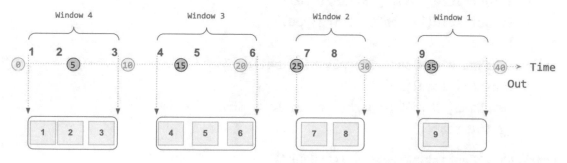

Figure 9-8. *Ten-second session window with a gap interval of five seconds*

Session windows are the right tools when business questions like which patients visited the hospital at a certain moment in time? Or what are the hospital busiest moments along a defined period of time?

As usual, we include a practical example of session window usage. In the following code snippet, you can see how session windows could be depicted as creating a window to collect all upcoming events arriving within the timeout period. As you see, all collected events inside the window time frame are added to the current session.

In the next example, we have implemented the session_window() to count incoming events over a session window with a ten-second gap on the Fecha column of our sample events.

```
// Session Window

import org.apache.spark.sql.SparkSession
import org.apache.spark.sql.functions._
import org.apache.spark.sql.types._
import org.apache.spark.sql.{DataFrame, Dataset, Encoders, SparkSession}
import java.io.IOException
import org.apache.spark.sql.streaming._
import org.apache.spark.sql.streaming.{GroupState,GroupStateTimeout,
OutputMode}
import org.apache.spark.sql.DataFrame

val PatientsSchema = StructType(Array(
      StructField("NSS", StringType),
      StructField("Nom", StringType),
      StructField("DID", IntegerType),
      StructField("DNom", StringType),
      StructField("Fecha", StringType))
      )
val spark:SparkSession = SparkSession.builder()
      .master("local[10]")
      .appName("Hand-On-Spark3_Socket_Data_Source")
      .getOrCreate()

spark.sparkContext.setLogLevel("ERROR")

import spark.implicits._
```

```
try {
      val PatientDS = spark.readStream
      .schema(PatientsSchema)
      .json("/tmp/window")

      PatientDS.printSchema()

      printf("\n Listening and ready... \n")

      val PatientDF = PatientDS
      .groupBy(
            session_window(col("Fecha"), "10 seconds"), col("DID")
      ).count()

      PatientDF.printSchema()

      PatientDF.writeStream
      .outputMode("complete")
      .format("console")
      .option("truncate", false)
      .start()
      .awaitTermination()

} catch {
      case e: IOException => println("IOException occurred")
      case t: Throwable => println("Error receiving data", t)
}finally {
      println("In finally block")
}
```

As you did in previous examples, before running the precedent code, you first have to create the data source folder (again "/tmp/window" in our example). After that you can copy JSON files provided as an example to the data source directory, for example, like this:

```
$ cp json_file11.json /tmp/window
$ cp json_file9.json /tmp/window
...
$ cp json_file7.json /tmp/window
```

Once the files are copied, your program should output something like this:

```
. . . Removed for brevity . . .

 Listening and ready...
----------------------------------------------
Batch: 0
----------------------------------------------

+---------------------------------------------------+---+-----+
|session_window                                     |DID|count|
+---------------------------------------------------+---+-----+
|{2023-02-23 01:00:15.002, 2023-02-23 01:00:25.002}|20 |1    |
|{2023-02-23 01:00:18.002, 2023-02-23 01:00:31.002}|10 |4    |
|{2023-02-23 01:00:17.002, 2023-02-23 01:00:27.002}|50 |1    |
|{2023-02-23 01:00:16.002, 2023-02-23 01:00:26.002}|30 |1    |
+---------------------------------------------------+---+-----+

----------------------------------------------
Batch: 1
----------------------------------------------

+---------------------------------------------------+---+-----+
|session_window                                     |DID|count|
+---------------------------------------------------+---+-----+
|{2023-02-23 01:00:15.002, 2023-02-23 01:00:25.002}|20 |2    |
|{2023-02-23 01:00:18.002, 2023-02-23 01:00:31.002}|10 |7    |
|{2023-02-23 01:00:17.002, 2023-02-23 01:00:27.002}|50 |2    |
|{2023-02-23 01:00:16.002, 2023-02-23 01:00:26.002}|30 |2    |
+---------------------------------------------------+---+-----+

----------------------------------------------
Batch: 2
----------------------------------------------

+---------------------------------------------------+---+-----+
|session_window                                     |DID|count|
+---------------------------------------------------+---+-----+
|{2023-02-23 01:02:00.002, 2023-02-23 01:02:10.002}|20 |1    |
|{2023-02-23 01:00:15.002, 2023-02-23 01:00:25.002}|20 |2    |
```

```
|{2023-02-23 01:01:34.002, 2023-02-23 01:01:44.002}|20 |1    |
|{2023-02-23 01:02:05.002, 2023-02-23 01:02:15.002}|10 |1    |
|{2023-02-23 01:00:18.002, 2023-02-23 01:00:31.002}|10 |7    |
|{2023-02-23 01:00:17.002, 2023-02-23 01:00:27.002}|50 |2    |
|{2023-02-23 01:02:20.002, 2023-02-23 01:02:30.002}|50 |1    |
|{2023-02-23 01:02:38.002, 2023-02-23 01:02:48.002}|50 |1    |
|{2023-02-23 01:01:30.002, 2023-02-23 01:01:43.002}|50 |4    |
|{2023-02-23 01:02:37.002, 2023-02-23 01:02:47.002}|30 |1    |
|{2023-02-23 01:00:16.002, 2023-02-23 01:00:26.002}|30 |2    |
|{2023-02-23 01:02:10.002, 2023-02-23 01:02:20.002}|30 |1    |
+------------------------------------------------+---+-----+


------------------------------------------
Batch: 3
------------------------------------------
+------------------------------------------------+---+-----+
|session_window                                  |DID|count|
+------------------------------------------------+---+-----+
|{2023-02-23 01:00:15.002, 2023-02-23 01:00:25.002}|20 |2    |
|{2023-02-23 01:01:34.002, 2023-02-23 01:01:44.002}|20 |1    |
|{2023-02-23 01:00:34.002, 2023-02-23 01:00:45.002}|20 |2    |
|{2023-02-23 01:02:00.002, 2023-02-23 01:02:10.002}|20 |1    |
|{2023-02-23 01:00:18.002, 2023-02-23 01:00:31.002}|10 |7    |
|{2023-02-23 01:02:05.002, 2023-02-23 01:02:15.002}|10 |1    |
|{2023-02-23 01:00:36.002, 2023-02-23 01:00:46.002}|10 |1    |
|{2023-02-23 01:02:20.002, 2023-02-23 01:02:30.002}|50 |1    |
|{2023-02-23 01:00:17.002, 2023-02-23 01:00:27.002}|50 |2    |
|{2023-02-23 01:00:30.002, 2023-02-23 01:00:48.002}|50 |5    |
|{2023-02-23 01:02:38.002, 2023-02-23 01:02:48.002}|50 |1    |
|{2023-02-23 01:01:30.002, 2023-02-23 01:01:43.002}|50 |4    |
|{2023-02-23 01:00:16.002, 2023-02-23 01:00:26.002}|30 |2    |
|{2023-02-23 01:02:10.002, 2023-02-23 01:02:20.002}|30 |1    |
|{2023-02-23 01:00:37.002, 2023-02-23 01:00:47.002}|30 |1    |
|{2023-02-23 01:02:37.002, 2023-02-23 01:02:47.002}|30 |1    |
+------------------------------------------------+---+-----+
```

Session Window with Dynamic Gap

Another interesting feature of the session window type is that it supports what is called *dynamic gap duration*[1]. The session window we implemented in our previous example, though it has a variable size depending on the arrival or not of new events, has a maximum size, the window length. The dynamic gap duration has the peculiarity of having a different gap duration per session. Thus, instead of a fixed gap/timeout value, we can use an expression to dynamically set the window size, adapting it to the input data characteristics.

In the following is a practical example showing you how to implement a session window with dynamic gap duration.

```
// Session Windows II. Session window with dynamic gap duration

import org.apache.spark.sql.SparkSession
import org.apache.spark.sql.functions._
import org.apache.spark.sql.types._
import org.apache.spark.sql.{DataFrame, Dataset, Encoders, SparkSession}
import java.io.IOException
import org.apache.spark.sql.streaming._
import org.apache.spark.sql.streaming.{GroupState,GroupStateTimeout,
OutputMode}
import org.apache.spark.sql.DataFrame

val PatientsSchema = StructType(Array(
    StructField("NSS", StringType),
    StructField("Nom", StringType),
    StructField("DID", IntegerType),
    StructField("DNom", StringType),
    StructField("Fecha", StringType))
    )
val spark:SparkSession = SparkSession.builder()
    .master("local[10]")
    .appName("Hand-On-Spark3_Socket_Data_Source")
    .getOrCreate()
```

---

[1] More information: www.databricks.com/blog/2021/10/12/native-support-of-session-window-in-spark-structured-streaming.html

```scala
spark.sparkContext.setLogLevel("ERROR")

import spark.implicits._

try {
      val PatientDS = spark.readStream
      .schema(PatientsSchema)
      .json("/tmp/window")

      PatientDS.printSchema()

      printf("\n Listening and ready... \n")

      val PatientDF = PatientDS
        .groupBy(
            session_window(col("Fecha"),
                      when(col("NSS") === "1009", "10 seconds")
                      .when(col("NSS") === "2001", "30 seconds")
                      .when(col("NSS") === "5000", "50 seconds")
                      .otherwise("60 seconds")),
            col("DID")
        ).count()

      PatientDF.printSchema()

      PatientDF.writeStream
      .outputMode("complete")
      .format("console")
      .option("truncate", false)
      .start()
      .awaitTermination()

} catch {
      case e: IOException => println("IOException occurred")
      case t: Throwable => println("Error receiving data", t)
}finally {
      println("In finally block")
}
```

The novelty of the previous code resides in this block of code, in which to implement the session window with a dynamic timeout.

```
val PatientDF = PatientDS
  .groupBy(
    session_window(col("Fecha"),
              when(col("NSS") === "1009", "10 seconds")
              .when(col("NSS") === "2001", "30 seconds")
              .when(col("NSS") === "5000", "50 seconds")
              .otherwise("60 seconds")),
      col("DID")
  ).count()
```

Now you can see that the session window length is dynamically established by the value of the NSS field.

Once more, after creating the source data directory ("/tmp/window") ,executing the code example, and copying the JSON files provided as examples, the program will get you an output similar to the next one.

```
. . . Removed for brevity . . .

 Listening and ready...

-------------------------------------------
Batch: 0
-------------------------------------------
+---------------------------------------------------+---+-----+
|session_window                                     |DID|count|
+---------------------------------------------------+---+-----+
|{2023-02-23 01:00:15.002, 2023-02-23 01:00:25.002}|20 |1    |
|{2023-02-23 01:00:18.002, 2023-02-23 01:01:20.002}|10 |3    |
|{2023-02-23 01:00:17.002, 2023-02-23 01:01:17.002}|50 |1    |
|{2023-02-23 01:00:16.002, 2023-02-23 01:01:16.002}|30 |1    |
+---------------------------------------------------+---+-----+
```

```
-----------------------------------------------
Batch: 1
-----------------------------------------------
```

session_window	DID	count
{2023-02-23 01:00:15.002, 2023-02-23 01:00:45.002}	20	2
{2023-02-23 01:00:18.002, 2023-02-23 01:01:21.002}	10	7
{2023-02-23 01:00:17.002, 2023-02-23 01:01:17.002}	50	2
{2023-02-23 01:00:16.002, 2023-02-23 01:01:16.002}	30	2

```
-----------------------------------------------
Batch: 2
-----------------------------------------------
```

session_window	DID	count
{2023-02-23 01:00:15.002, 2023-02-23 01:01:22.002}	20	3
{2023-02-23 01:00:18.002, 2023-02-23 01:01:28.002}	10	11
{2023-02-23 01:00:17.002, 2023-02-23 01:01:29.002}	50	4
{2023-02-23 01:00:16.002, 2023-02-23 01:01:23.002}	30	3

```
-----------------------------------------------
Batch: 3
-----------------------------------------------
```

session_window	DID	count
{2023-02-23 01:00:15.002, 2023-02-23 01:01:35.002}	20	5
{2023-02-23 01:00:18.002, 2023-02-23 01:01:36.002}	10	12
{2023-02-23 01:00:17.002, 2023-02-23 01:01:38.002}	50	9
{2023-02-23 01:00:16.002, 2023-02-23 01:01:37.002}	30	4

```
-------------------------------------------
Batch: 4
-------------------------------------------
+--------------------------------------------------+---+-----+
|session_window                                    |DID|count|
+--------------------------------------------------+---+-----+
|{2023-02-23 01:00:15.002, 2023-02-23 01:03:00.002}|20 |7    |
|{2023-02-23 01:02:05.002, 2023-02-23 01:03:05.002}|10 |1    |
|{2023-02-23 01:00:18.002, 2023-02-23 01:01:36.002}|10 |12   |
|{2023-02-23 01:00:17.002, 2023-02-23 01:03:38.002}|50 |15   |
|{2023-02-23 01:00:16.002, 2023-02-23 01:01:37.002}|30 |4    |
|{2023-02-23 01:02:10.002, 2023-02-23 01:03:37.002}|30 |2    |
+--------------------------------------------------+---+-----+


-------------------------------------------
Batch: 5
-------------------------------------------
+--------------------------------------------------+---+-----+
|session_window                                    |DID|count|
+--------------------------------------------------+---+-----+
|{2023-02-23 01:00:15.002, 2023-02-23 01:03:00.002}|20 |7    |
|{2023-02-23 01:00:18.002, 2023-02-23 01:01:36.002}|10 |12   |
|{2023-02-23 01:02:05.002, 2023-02-23 01:03:05.002}|10 |1    |
|{2023-02-23 01:00:17.002, 2023-02-23 01:03:38.002}|50 |15   |
|{2023-02-23 01:02:10.002, 2023-02-23 01:03:37.002}|30 |2    |
|{2023-02-23 01:00:10.002, 2023-02-23 01:01:37.002}|30 |5    |
+--------------------------------------------------+---+-----+
```

At the time this book was written and as for Spark 3.3.2, some restrictions are in place when using session windows in streaming query:

- Output mode "update" is not supported.

- The grouping clause should include at least two columns, the session_window and another one.

However, when used for batch query, grouping clauses can only include the session_ window column as mentioned in the Apache Spark official documentation[2].

9.6 Watermarking in Spark Structured Streaming

As we have already mentioned several times across this book, stream data is far from ideal. We have already gone through the concepts of stranglers and late-arrival events.

Watermarking was introduced in Apache Spark 2.1 to support late-arriving data. For example, watermarks are used in stateful streaming operations to avoid boundlessly accumulating information in state, which in all likelihood will provoke instability due to memory saturation, hence introducing computing latencies in the course of streaming operations.

This section explains the basic concepts behind Watermarking and provides a practical example for using watermarks with Spark stateful streaming operations.

What Is a Watermark?

Watermarking could be defined as a lateness threshold. Watermarking permits Spark Structured Streaming to tackle the problem of late-arrival events. Management of stragglers or out-of-order events is critical in distributed architectures for the sake of data integrity, accuracy, and fault tolerance. When dealing with this kind of complex system, it is not guaranteed that the data will arrive to the streaming platform in the order it was delivered. This could happen due to network bottlenecks, latency in the communications, etc. To overcome these difficulties, the state of aggregate operations must be retained.

Spark Structured Streaming uses watermarks as a cutoff point to control for how long the Spark Stream Processing Engine will wait for late events.

Therefore, when we declare a watermark, we specify a timestamp field and a watermark limit of time. For instance, consider our Session Windows code snippet. We can modify it as shown in the following, to introduce a watermark threshold.

---

[2]https://spark.apache.org/docs/latest/structured-streaming-programming-guide. html#types-of-time-windows

```
val PatientDF = PatientDS
.withWatermark("Fecha", "30 seconds")
.groupBy(
        session_window(col("Fecha"), "10 seconds"), col("DID")
).count()
```

In this example

- The Fecha column is used to define a 30 seconds' watermark.

- A count is performed for each DID observed for each nonoverlapping ten seconds' window.

- State information is preserved for each count until the end of the window is ten seconds older than the latest observed Fecha value.

After including a watermark, as new data arrives, Spark tracks the most recent timestamp in the designated column and processes the incoming event within the watermark threshold.

Here is the complete code example including a watermark of 30 seconds.

```
// Watermarking in Spark Structured Streaming

import org.apache.spark.sql.SparkSession
import org.apache.spark.sql.functions._
import org.apache.spark.sql.types._ //{IntegerType, StringType,
StructField, StructType,DoubleType,LongType}
import org.apache.spark.sql.{DataFrame, Dataset, Encoders, SparkSession}
import java.io.IOException
import org.apache.spark.sql.streaming._
import org.apache.spark.sql.streaming.{GroupState,GroupStateTimeout,
OutputMode}
import org.apache.spark.sql.DataFrame

val PatientsSchema = StructType(Array(
     StructField("NSS", StringType),
     StructField("Nom", StringType),
     StructField("DID", IntegerType),
```

```scala
      StructField("DNom", StringType),
      StructField("Fecha", StringType))
      )
val spark:SparkSession = SparkSession.builder()
      .master("local[10]")
      .appName("Hand-On-Spark3_Socket_Data_Source")
      .getOrCreate()

spark.sparkContext.setLogLevel("ERROR")

import spark.implicits._

try {
      val PatientDS = spark.readStream
      .schema(PatientsSchema)
      .json("/tmp/window")
      .withColumn("Fecha", to_timestamp(col("Fecha"),
      "yyyy-MM-dd'T'HH:mm:ss.SSSX"))

      PatientDS.printSchema()

      printf("\n Listening and ready... \n")

      val PatientDF = PatientDS
        .withWatermark("Fecha", "30 seconds")
          .groupBy(
            session_window(col("Fecha"), "10 seconds"), col("DID")
        ).count()

      PatientDF.printSchema()

      PatientDF.writeStream
      .outputMode("complete")
      .format("console")
      .option("truncate", false)
      .start()
      .awaitTermination()
```

```
} catch {
     case e: IOException => println("IOException occurred")
     case t: Throwable => println("Error receiving data", t)
}finally {
     println("In finally block")
}
```

There is another important part of the precedent code snippet you should pay attention to.

```
val PatientDS = spark.readStream
.schema(PatientsSchema)
.json("/tmp/window")
.withColumn("Fecha", to_timestamp(col("Fecha"),
"yyyy-MM-dd'T'HH:mm:ss.SSSX"))

PatientDS.printSchema()
```

```
root
 |-- NSS: string (nullable = true)
 |-- Nom: string (nullable = true)
 |-- DID: integer (nullable = true)
 |-- DNom: string (nullable = true)
 |-- Fecha: timestamp (nullable = true)
```

Watermark can only be used with timestamp or window columns. Thus, column Fecha must be converted from string to timestamp type before it can be used; otherwise, you will get an error.

Once again, if you run the precedent program and copy the JSON files provided as examples to the data source directory ("/tmp/window"), you will get an output similar to the following one.

```
. . . Removed for brevity . . .

 Listening and ready...
-------------------------------------------
Batch: 0
-------------------------------------------
```

```
+----------------------------------------------------+---+-----+
|session_window                                      |DID|count|
+----------------------------------------------------+---+-----+
|{2023-02-23 01:01:34.002, 2023-02-23 01:01:44.002}|20 |1    |
|{2023-02-23 01:02:00.002, 2023-02-23 01:02:10.002}|20 |1    |
|{2023-02-23 01:02:05.002, 2023-02-23 01:02:15.002}|10 |1    |
|{2023-02-23 01:02:20.002, 2023-02-23 01:02:30.002}|50 |1    |
|{2023-02-23 01:02:38.002, 2023-02-23 01:02:48.002}|50 |1    |
|{2023-02-23 01:01:30.002, 2023-02-23 01:01:43.002}|50 |4    |
|{2023-02-23 01:02:10.002, 2023-02-23 01:02:20.002}|30 |1    |
|{2023-02-23 01:02:37.002, 2023-02-23 01:02:47.002}|30 |1    |
+----------------------------------------------------+---+-----+


-------------------------------------------
Batch: 1
-------------------------------------------

+----------------------------------------------------+---+-----+
|session_window                                      |DID|count|
+----------------------------------------------------+---+-----+
|{2023-02-23 01:02:00.002, 2023-02-23 01:02:10.002}|20 |1    |
|{2023-02-23 01:01:34.002, 2023-02-23 01:01:44.002}|20 |1    |
|{2023-02-23 01:02:05.002, 2023-02-23 01:02:15.002}|10 |1    |
|{2023-02-23 01:02:20.002, 2023-02-23 01:02:30.002}|50 |1    |
|{2023-02-23 01:02:38.002, 2023-02-23 01:02:48.002}|50 |1    |
|{2023-02-23 01:01:30.002, 2023-02-23 01:01:43.002}|50 |4    |
|{2023-02-23 01:02:37.002, 2023-02-23 01:02:47.002}|30 |1    |
|{2023-02-23 01:02:10.002, 2023-02-23 01:02:20.002}|30 |1    |
+----------------------------------------------------+---+-----+


-------------------------------------------
Batch: 2
-------------------------------------------

+----------------------------------------------------+---+-----+
|session_window                                      |DID|count|
+----------------------------------------------------+---+-----+
|{2023-02-23 01:02:00.002, 2023-02-23 01:02:10.002}|20 |1    |
|{2023-02-23 01:01:34.002, 2023-02-23 01:01:44.002}|20 |1    |
```

```
|{2023-02-23 01:02:05.002, 2023-02-23 01:02:15.002}|10 |1    |
|{2023-02-23 01:02:20.002, 2023-02-23 01:02:30.002}|50 |1    |
|{2023-02-23 01:02:38.002, 2023-02-23 01:02:48.002}|50 |1    |
|{2023-02-23 01:01:30.002, 2023-02-23 01:01:43.002}|50 |4    |
|{2023-02-23 01:02:37.002, 2023-02-23 01:02:47.002}|30 |1    |
|{2023-02-23 01:02:10.002, 2023-02-23 01:02:20.002}|30 |1    |
+---------------------------------------------------+---+-----+


-------------------------------------------
Batch: 3
-------------------------------------------
+---------------------------------------------------+---+-----+
|session_window                                     |DID|count|
+---------------------------------------------------+---+-----+
|{2023-02-23 01:02:00.002, 2023-02-23 01:02:10.002}|20 |1    |
|{2023-02-23 01:01:34.002, 2023-02-23 01:01:44.002}|20 |1    |
|{2023-02-23 01:02:05.002, 2023-02-23 01:02:15.002}|10 |1    |
|{2023-02-23 01:02:20.002, 2023-02-23 01:02:30.002}|50 |1    |
|{2023-02-23 01:02:38.002, 2023-02-23 01:02:48.002}|50 |1    |
|{2023-02-23 01:01:30.002, 2023-02-23 01:01:43.002}|50 |4    |
|{2023-02-23 01:02:37.002, 2023-02-23 01:02:47.002}|30 |1    |
|{2023-02-23 01:02:10.002, 2023-02-23 01:02:20.002}|30 |1    |
+---------------------------------------------------+---+-----+


-------------------------------------------
Batch: 4
-------------------------------------------
+---------------------------------------------------+---+-----+
|session_window                                     |DID|count|
+---------------------------------------------------+---+-----+
|{2023-02-23 01:02:00.002, 2023-02-23 01:02:10.002}|20 |1    |
|{2023-02-23 01:01:34.002, 2023-02-23 01:01:44.002}|20 |1    |
|{2023-02-23 01:02:05.002, 2023-02-23 01:02:15.002}|10 |1    |
|{2023-02-23 01:02:20.002, 2023-02-23 01:02:30.002}|50 |1    |
|{2023-02-23 01:02:38.002, 2023-02-23 01:02:48.002}|50 |1    |
|{2023-02-23 01:01:30.002, 2023-02-23 01:01:43.002}|50 |4    |
```

```
|{2023-02-23 01:02:37.002, 2023-02-23 01:02:47.002}|30 |1    |
|{2023-02-23 01:02:10.002, 2023-02-23 01:02:20.002}|30 |1    |
+--------------------------------------------------+---+-----+


-------------------------------------------
Batch: 5
-------------------------------------------

+--------------------------------------------------+---+-----+
|session_window                                    |DID|count|
+--------------------------------------------------+---+-----+
|{2023-02-23 01:02:00.002, 2023-02-23 01:02:10.002}|20 |1    |
|{2023-02-23 01:01:34.002, 2023-02-23 01:01:44.002}|20 |1    |
|{2023-02-23 01:02:05.002, 2023-02-23 01:02:15.002}|10 |1    |
|{2023-02-23 01:02:20.002, 2023-02-23 01:02:30.002}|50 |1    |
|{2023-02-23 01:02:38.002, 2023-02-23 01:02:48.002}|50 |1    |
|{2023-02-23 01:01:30.002, 2023-02-23 01:01:43.002}|50 |4    |
|{2023-02-23 01:02:37.002, 2023-02-23 01:02:47.002}|30 |1    |
|{2023-02-23 01:02:10.002, 2023-02-23 01:02:20.002}|30 |1    |
+--------------------------------------------------+---+-----+
```

9.7 Summary

In this chapter, we covered the different Event-Time Window Operations and Watermarking with Apache Spark. First, we studied how to perform streaming aggregations with the tumbling and sliding windows, the two types of fixed-sized window operations. After that we learned how to implement a session window and how to use the new Spark built-in function session_window to create a window column. Special attention was paid to the session window with dynamic gap duration to adapt the window length as a function of the input data. Finally, we have covered Watermarking in Spark Structured Streaming and how it can be used to manage late-arriving events. In the next and final chapter, we are going to explore future directions for Spark Streaming.

CHAPTER 10

Future Directions for Spark Streaming

Nowadays, data drives many decision-making processes for companies worldwide. Information assists them in understanding their customers and attracting new ones. Information is also used to streamline business processes and achieve both competitive advantage and operational efficiency. These are the reasons why so many companies understand the importance of data to make better decisions, improve customer relationships, and launch strategic initiatives. Therefore, to take full advantage of information, companies have to know how to extract its value.

At the same time, as the world digitizes, data is increasingly considered a depreciating asset because organizations need to gain insights of huge volumes of information in near real time. That is where stream processing comes in.

We have seen in previous chapters that stream processing permits processing and analyzing live information. We have also seen that real-time information processing can be used to instantly react to events and trigger proactive or reactive actions. Thus, stream analysis enables businesses and organizations to take immediate action on opportunities and threats.

Stream processing is finding increasing applications in event-driven architectures and microservice processes orchestration. Another use of stream processing gaining popularity nowadays is by coupling real-time data processing with artificial intelligence (AI) and machine learning (ML) algorithms to make instantaneous predictions.

In the next section, we are going to show you a practical example of how to use Spark Structured Streaming with Spark ML to apply ML algorithms to data streams, extract patterns, gain insights into live information, and trigger decisions.

© Alfonso Antolínez García 2023
A. Antolínez García, *Hands-on Guide to Apache Spark 3*, https://doi.org/10.1007/978-1-4842-9380-5_10

10.1 Streaming Machine Learning with Spark

As we have already mentioned in this chapter, many data sources produce data in an unbounded manner, for example, web logs, Internet of things (IoT) devices, transactions from financial services, etc. These continuous streams of data were traditionally stored and converted into bounded datasets for later batch processing. Therefore, data collection, processing, and preparation for decision-making were complete asynchronous processes, occurring at different moments in time.

In our time, many organizations simply cannot afford such a time delay between data intake and decision-making due to the time value of the data. These days, many scenarios require taking advantage of live data to proactively respond to events as close as the information is available. For instance, consider use cases such as autonomous driving, unmanned vehicles, etc.

In circumstances like that, rather than wait for the data to go through the whole process, streaming analytics permits the detection of patterns in data in almost real time and consequently triggers actions. Another important advantage of performing analyses of in motion data is that as information properties and its patterns change over time, streaming algorithms can adapt to them.

This section introduces how Spark Machine Learning (Spark ML) and Spark Streaming can be coupled together to make predictions on streaming data.

Next code example shows how to train a machine learning model using Spark ML to generate a `PipelineModel` to make predictions on streaming workflows. It is out of scope of this book to teach you how to implement machine learning with Spark; thus, it is assumed the reader has a basic knowledge of machine learning and how to implement it with Apache Spark ML.

For the purpose of this example, we are going to use a small dataset of 303 rows and 14 columns for heart attack classification that is available for public domain and which can be found at Heart Attack Analysis & Prediction Dataset. After training or ML model on this dataset, we will be able to predict whether a person can suffer a heart attack based on 13 dependent variables such as age, sex, and vital signs.

Table 10-1 shows a description of each column of the dataset.

Table 10-1. *Heart Attack Analysis and Prediction Dataset Columns Description*

Column name	Description and possible values	
age	Age of the patient	
sex	Sex of the patient	
exang	exercise induced angina	
	Value	
	1	Yes
	0	No
ca	Number of major vessels (0–3)	
cp	Chest pain type	
	Value	
	1	Typical angina
	2	Atypical angina
	3	Non-anginal pain
	4	Asymptomatic
trtbps	Resting blood pressure (in mm Hg)	
chol	Cholesterol in mg/dl fetched via BMI sensor	
fbs	(Fasting blood sugar > 120 mg/dl) (1 = true; 0 = false)	
rest_ecg:	Resting electrocardiographic results	
	Value	
	0	Normal
	1	Having ST-T wave abnormality (T wave inversions and/or ST elevation or depression of > 0.05 mV)

(continued)

Table 10-1. (*continued*)

Column name	Description and possible values	
2		Showing probable or definite left ventricular hypertrophy by Estes' criteria
thalach	Maximum heart rate achieved	
output		
	Value	
	0	Less chance of heart attack
	1	More chance of heart attack

The column "output" represents the dependent variable, and as you can see, it can only take two possible values: 0 and 1. Therefore, we have to deal with a binary classification problem. For that reason, we can implement a logistic regression model as it is suitable for probability prediction.

What Is Logistic Regression?

Regression model (also known as logit model) is commonly used for classification and predictive analytics. Logistic regression estimates the probability of an event occurring, such as heart attack or no heart attack, based on a given database of independent variables also called predictors.

Logistic regression is used to estimate the relationship between a dependent and continuous variable and one or more independent categorical variables. Categorical variables can only acquire a limited number of values, that is, true or false, yes or no, 1 or 0, etc.

Under the context of machine learning, logistic regression belongs to the family of supervised machine learning models. Supervised machine learning models require a labeled dataset to train the model.

Types of Logistic Regression

There are three types of logistic regression models based on their categorical output.

- Binary logistic regression: The dependent variable has only two possible outcomes (e.g., 0 or 1).

- Multinomial logistic regression: The dependent variable has three or more possible outcomes and these values have no specified order.

- Ordinal logistic regression: The dependent variable has three or more possible outcomes, and these values have a specific order.

Use Cases of Logistic Regression

Logistic regression can be used for regression (prediction) and classification problems. Some of these use cases could be the following:

- Fraud detection: Identification of anomalies which are predictive of fraud

- Disease prediction: In medicine, prediction of the likelihood of a disease for a given population

Thus, logistic regression can facilitate prediction and enhance decision-making. More information about the logistic regression can be found here.

After this short introduction about the concepts behind logistic regression, let's now focus on our code example.

First of all, we are going to create a schema for our dataframe to enumerate the columns and their types of data while loading the data.

Here is the code.

```
import org.apache.spark.sql.types.{StructType,LongType}
import org.apache.spark.ml.feature.{OneHotEncoder, VectorAssembler,
MinMaxScaler, StringIndexer}
import org.apache.spark.ml.{Pipeline, PipelineModel}
import org.apache.spark.ml.classification.LogisticRegression

val schema = new StructType()
      .add("age",LongType,true)
      .add("sex",LongType,true)
```

```scala
        .add("cp",LongType,true)
        .add("trtbps",LongType,true)
        .add("chol",LongType,true)
        .add("fbs",LongType,true)
        .add("restecg",LongType,true)
        .add("thalachh",LongType,true)
        .add("exng",LongType,true)
        .add("oldpeak",LongType,true)
        .add("slp",LongType,true)
        .add("caa",LongType,true)
        .add("thall",LongType,true)
        .add("output",LongType,true)

val spark:SparkSession = SparkSession.builder()
    .master("local[*]")
    .appName("Hand-On-Spark3_Spark_ML_and_Streaming")
    .getOrCreate()

spark.sparkContext.setLogLevel("ERROR")

val heartdF = spark.read.format("csv")
      .option("header", "true")
      .schema(schema)
      .load("file:///tmp/spark_ml")
      .withColumnRenamed("output","label")

println(heartdF.count)
heartdF.printSchema()
```

When you run this code, for instance, in a notebook, you will find the following output.

```
303
root
 |-- age: long (nullable = true)
 |-- sex: long (nullable = true)
 |-- cp: long (nullable = true)
 |-- trtbps: long (nullable = true)
 |-- chol: long (nullable = true)
```

```
|-- fbs: long (nullable = true)
|-- restecg: long (nullable = true)
|-- thalachh: long (nullable = true)
|-- exng: long (nullable = true)
|-- oldpeak: long (nullable = true)
|-- slp: long (nullable = true)
|-- caa: long (nullable = true)
|-- thall: long (nullable = true)
|-- label: long (nullable = true)
```

You can see in the preceding code the schema of the dataframe and column's data types.

A very important step when working with data is the process of data engineering and feature engineering. As part of the data engineering process, it is always recommended to check the existence of NULL values in our dataset.

If inadvertently you process a dataset with NULL values, at best you will receive an error and understand something is wrong with the data, and at worst you will get inaccurate results.

In our dataset, if you check the "oldpeak" column, running the following line of code, you will find there are 173 NULL values

```
heartdF.filter("oldpeak is null").count
  res2: Long = 173
```

Therefore, we will have to take it into consideration along the construction of our ML model.

The next step could be the split of our dataset between training (trainDF) and test (testDF) subdatasets.

```
val Array(trainDF, testDF) = heartdF.randomSplit(weights=Array(.8, .2))
```

The previous line of code will randomly split the data in a 80%–20% proportion. Eighty percent of the data will be used to train our PipelineModel and the other 20% (the unseen data) to test it.

```
val lr = new LogisticRegression()
  .setMaxIter(10)
  .setRegParam(0.01)
```

```
val oneHotEnc = new OneHotEncoder()
.setInputCols(Array("sex", "cp", "fbs", "restecg", "exng", "slp",
"caa","thall"))
.setOutputCols(Array("SexOHE", "cpOHE", "fbsOHE", "restecgOHE", "exngOHE",
"slpOHE", "caaOHE","thallOHE"))

val assemblerA = new VectorAssembler()
  .setInputCols(Array("age", "trtbps", "chol", "thalachh", "oldpeak"))
  .setOutputCol("features_scaled1")
  .setHandleInvalid("skip")

val scaler = new MinMaxScaler()
  .setInputCol("features_scaled1")
  .setOutputCol("features_scaled")

val assemblerB = new VectorAssembler()
  .setInputCols(Array("SexOHE", "cpOHE", "fbsOHE", "restecgOHE", "exngOHE",
  "slpOHE", "caaOHE","thallOHE", "features_scaled"))
  .setOutputCol("features")
  .setHandleInvalid("skip")

val modelStages = Array(assemblerA, scaler, oneHotEnc, assemblerB, lr)

val pipeline = new Pipeline()
  .setStages(modelStages)

val PipelineModel = pipeline.fit(trainDF)

val trainingPred = PipelineModel.transform(trainDF)

trainingPred.select("label","probability","prediction").
show(truncate=false)
```

If you execute the precedent piece of code in your notebook, you will get an output pretty similar to the next one.

```
+-----+---------------------------------------------+----------+
|label|probability                                  |prediction|
+-----+---------------------------------------------+----------+
|1    |[0.03400091691592197,0.965999083084078]      |1.0       |
|1    |[0.05511659822191829,0.9448834017780817]     |1.0       |
|0    |[0.5605994301074364,0.4394005698925636]      |0.0       |
|1    |[0.03115074381750154,0.9688492561824985]     |1.0       |
|1    |[0.004384634167846924,0.995615365832153]     |1.0       |
|1    |[0.08773404036960819,0.9122659596303918]     |1.0       |
|1    |[0.08773404036960819,0.9122659596303918]     |1.0       |
|1    |[0.06985863429068614,0.9301413657093138]     |1.0       |
|0    |[0.7286457381073151,0.27135426189268486]     |0.0       |
|1    |[0.02996587703476992,0.9700341229652301]     |1.0       |
|1    |[0.0016700146317826447,0.9983299853682174]   |1.0       |
|0    |[0.36683434534535186,0.6331656546546481]     |1.0       |
|1    |[0.04507024193962369,0.9549297580603763]     |1.0       |
|1    |[0.013996165515300337,0.9860038344846996]    |1.0       |
|1    |[0.016828318827434772,0.9831716811725653]    |1.0       |
|1    |[0.2671331307894787,0.7328668692105214]      |1.0       |
|1    |[0.32331781956753536,0.6766821804324646]     |1.0       |
|1    |[0.09759145569985764,0.9024085443001424]     |1.0       |
|1    |[0.032829375720753985,0.967170624279246]     |1.0       |
|0    |[0.8584162531850159,0.1415837468149841]      |0.0       |
+-----+---------------------------------------------+----------+
only showing top 20 rows
```

If you pay attention to the previous outcome, you will see that the lower the probability, the more likely the prediction to be 1, while on the other hand, the higher the probability, the more likely the prediction to be 0.

One line of the previous code you should pay attention to is this one:

```
.setHandleInvalid("skip")
```

If you remember, our dataset has columns with NULL values. If you do not take care of them, you will receive an error. The previous line of code skips NULL values.

Once we have trained our model, we are going to divide our test dataset (tetDF) into multiple files to simulate a streaming data flow. Then, we are going to set up a file data source and copy each individual file to the source folder, as we did in previous chapters simulating a stream of information.

Next is the code to divide testDF into ten partitions (individual files) and writing them to the /tmp/spark_ml_streaming/ directory.

```
testDF.repartition(10)
.write.format("csv")
.option("header", true)
.mode("overwrite")
.save("file:///tmp/spark_ml_streaming/")
```

After executing the previous code snippet, if you have a look at the designated source directory, you will find something similar to this:

```
$ tree /tmp/spark_ml_streaming/
/tmp/spark_ml_streaming/
├── part-00000-2c24d64a-2ecd-4674-a394-44aa5e17f131-c000.csv
├── part-00001-2c24d64a-2ecd-4674-a394-44aa5e17f131-c000.csv
├── part-00002-2c24d64a-2ecd-4674-a394-44aa5e17f131-c000.csv
├── part-00003-2c24d64a-2ecd-4674-a394-44aa5e17f131-c000.csv
├── part-00004-2c24d64a-2ecd-4674-a394-44aa5e17f131-c000.csv
├── part-00005-2c24d64a-2ecd-4674-a394-44aa5e17f131-c000.csv
├── part-00006-2c24d64a-2ecd-4674-a394-44aa5e17f131-c000.csv
├── part-00007-2c24d64a-2ecd-4674-a394-44aa5e17f131-c000.csv
├── part-00008-2c24d64a-2ecd-4674-a394-44aa5e17f131-c000.csv
├── part-00009-2c24d64a-2ecd-4674-a394-44aa5e17f131-c000.csv
└── _SUCCESS
```

Next, we have to create the streaming source to load the files from the data source as soon as they appear in the directory.

```
val streamingSource=spark
    .readStream
    .format("csv")
    .option("header",true)
    .schema(schema)
```

```
.option("ignoreLeadingWhiteSpace",true)
.option("mode","dropMalformed")
.option("maxFilesPerTrigger",1)
.load("file:///tmp/HeartTest/")
.withColumnRenamed("output","label")
```

We have to control the quality of the data that is injected into the model; that is why we have included the following lines:

```
.option("ignoreLeadingWhiteSpace",true)
.option("mode","dropMalformed")
```

to be sure that unnecessary white spaces and malformed rows do not get to the model.

We have also added the line

```
.option("maxFilesPerTrigger",1)
```

to be sure only one file is processed at a time.

It is now time to write our PipelineModel, pass the input stream through it, and construct the stream writer to pour the output into the sink.

```
val streamingHeart = PipelineModel.transform(streamingSource).select
("label","probability","prediction")

streamingHeart.writeStream
      .outputMode("append")
      .option("truncate", false)
      .format("console")
      .start()
      .awaitTermination()
```

Now, execute the precedent code snippet and copy the partitioned files to the data source. For example

```
$ cp part-00000-2c24d64a-2ecd-4674-a394-44aa5e17f131-c000.csv /tmp/HeartTest/
$ cp part-00001-2c24d64a-2ecd-4674-a394-44aa5e17f131-c000.csv /tmp/HeartTest/
...
$ cp part-00009-2c24d64a-2ecd-4674-a394-44aa5e17f131-c000.csv /tmp/HeartTest/
```

You will see an output similar to the next one, coming out of your program.

```
-------------------------------------------
Batch: 0
-------------------------------------------
+-----+-------------------------------------------+----------+
|label|probability                                |prediction|
+-----+-------------------------------------------+----------+
|0    |[0.7464870545074516,0.25351294549254844]   |0.0       |
|1    |[0.1632367041842738,0.8367632958157262]    |1.0       |
+-----+-------------------------------------------+----------+


-------------------------------------------
Batch: 1
-------------------------------------------
+-----+-------------------------------------------+----------+
|label|probability                                |prediction|
+-----+-------------------------------------------+----------+
|0    |[0.9951659487928823,0.004834051207117662]  |0.0       |
|0    |[0.9929886660069713,0.007011333993028668]  |0.0       |
+-----+-------------------------------------------+----------+


-------------------------------------------
Batch: 2
-------------------------------------------
+-----+-------------------------------------------+----------+
|label|probability                                |prediction|
+-----+-------------------------------------------+----------+
|0    |[0.6601488743972465,0.33985112560275355]   |0.0       |
|0    |[0.9885105583774811,0.011489441622518859]  |0.0       |
|1    |[0.004729033461790646,0.9952709665382093]  |1.0       |
|1    |[0.002543643876197849,0.9974563561238021]  |1.0       |
+-----+-------------------------------------------+----------+
```

```
-----------------------------------------------
Batch: 3
-----------------------------------------------
+-----+--------------------------------------+----------+
|label|probability                           |prediction|
+-----+--------------------------------------+----------+
|1    |[0.23870496408150266,0.7612950359184973]|1.0       |
|0    |[0.8285765606366566,0.17142343936334337]|0.0       |
|1    |[0.1123278992547269,0.8876721007452731] |1.0       |
+-----+--------------------------------------+----------+


-----------------------------------------------
Batch: 4
-----------------------------------------------
+-----+--------------------------------------+----------+
|label|probability                           |prediction|
+-----+--------------------------------------+----------+
|1    |[0.3811392681451562,0.6188607318548438] |1.0       |
|1    |[0.016044469761318698,0.9839555302386813]|1.0       |
|1    |[0.011124987326959632,0.9888750126730403]|1.0       |
|0    |[0.009425069592366693,0.9905749304076333]|1.0       |
+-----+--------------------------------------+----------+


-----------------------------------------------
Batch: 5
-----------------------------------------------
+-----+--------------------------------------+----------+
|label|probability                           |prediction|
+-----+--------------------------------------+----------+
|1    |[0.030581176663381764,0.9694188233366182]|1.0       |
|1    |[0.028952221072329157,0.9710477789276708]|1.0       |
|0    |[0.7251959061823547,0.27480409381764526] |0.0       |
+-----+--------------------------------------+----------+
```

```
------------------------------------------
Batch: 6
------------------------------------------
+-----+----------------------------------------+----------+
|label|probability                             |prediction|
+-----+----------------------------------------+----------+
|1    |[0.3242653848343221,0.6757346151656779] |1.0       |
|0    |[0.9101196538221397,0.08988034617786034]|0.0       |
|1    |[0.08227291309126751,0.9177270869087325]|1.0       |
+-----+----------------------------------------+----------+


------------------------------------------
Batch: 7
------------------------------------------
+-----+----------------------------------------+----------+
|label|probability                             |prediction|
+-----+----------------------------------------+----------+
|1    |[0.09475287521715883,0.9052471247828412]|1.0       |
+-----+----------------------------------------+----------+


------------------------------------------
Batch: 8
------------------------------------------
+-----+----------------------------------------+----------+
|label|probability                             |prediction|
+-----+----------------------------------------+----------+
|1    |[0.8256079035149502,0.17439209648504983]|0.0       |
|0    |[0.31539711793989017,0.6846028820601098]|1.0       |
|0    |[0.9889473486170233,0.01105265138297673]|0.0       |
|1    |[0.12416982209602322,0.8758301779039768]|1.0       |
+-----+----------------------------------------+----------+
```

When developing a machine learning (ML) model, it is always essential to find out whether it accurately measures what it is set out to measure.

In the next section, we are going to introduce a small variation in our example code to show you how to assess the accuracy of a pipeline model through the measure of its sensitivity and specificity.

Assessing the Sensitivity and Specificity of Our Streaming ML Model

As mentioned just previously, it is not enough that a ML model makes predictions; those predictions have to be accurate.

Sensitivity and specificity are metrics that indicate the accuracy of a test or measure and help to determine how valid the predictions are. Whenever we create a ML model, in this case to screen for the possibility of a person suffering a heart attack, or to detect an abnormality, we must determine how valid that model is. In this heart attack analysis and prediction example, our screening model is used to decide which patients are more likely to have a condition (heart attack).

Next we show you how you can also adapt the following part of the previous code:

```
val streamingHeart = PipelineModel.transform(streamingSource).select
("label","probability","prediction")

streamingHeart.writeStream
    .outputMode("append")
    .option("truncate", false)
    .format("console")
    .start()
    .awaitTermination()
```

changing it like this

```
import org.apache.spark.sql.functions.{count, sum, when}

val streamingRates = PipelineModel.transform(streamingSource)
    .groupBy('label)
    .agg(
    (sum(when('prediction === 'label, 1)) / count('label)).alias("true
    prediction rate"),
    count('label).alias("count")
    )

streamingRates.writeStream
    .outputMode("complete")
    .option("truncate", false)
```

```
        .format("console")
        .start()
        .awaitTermination()
```

to calculate the ongoing sensitivity and specificity, respectively, of the predictions of the model for the test dataset.

As we are applying our PipelineModel to a stream of data, the previous metrics (sensitivity and specificity) are going to be calculated as the rates of true positive and true negative predictions and constantly being updated as the incoming data is processed.

After adapting the code and repeating the streaming simulation process, your code will show you an output similar to the one shown in the following.

```
-----------------------------------------------
Batch: 0
-----------------------------------------------
+-----+--------------------+-----+
|label|true prediction rate|count|
+-----+--------------------+-----+
|0    |0.5                 |2    |
|1    |0.5                 |2    |
+-----+--------------------+-----+

-----------------------------------------------
Batch: 1
-----------------------------------------------
+-----+--------------------+-----+
|label|true prediction rate|count|
+-----+--------------------+-----+
|0    |0.6666666666666666  |3    |
|1    |0.6666666666666666  |3    |
+-----+--------------------+-----+

-----------------------------------------------
Batch: 2
-----------------------------------------------
```

```
+-----+-------------------+-----+
|label|true prediction rate|count|
+-----+-------------------+-----+
|0    |0.6666666666666666 |3    |
|1    |0.7142857142857143 |7    |
+-----+-------------------+-----+


-------------------------------------------
Batch: 3
-------------------------------------------
+-----+-------------------+-----+
|label|true prediction rate|count|
+-----+-------------------+-----+
|0    |0.75               |4    |
|1    |0.75               |8    |
+-----+-------------------+-----+


-------------------------------------------
Batch: 4
-------------------------------------------
+-----+-------------------+-----+
|label|true prediction rate|count|
+-----+-------------------+-----+
|0    |0.8                |5    |
|1    |0.7777777777777778 |9    |
+-----+-------------------+-----+


-------------------------------------------
Batch: 5
-------------------------------------------
+-----+-------------------+-----+
|label|true prediction rate|count|
+-----+-------------------+-----+
|0    |0.8                |5    |
|1    |0.6363636363636364 |11   |
+-----+-------------------+-----+
```

```
-------------------------------------------
Batch: 6
-------------------------------------------
+-----+--------------------+-----+
|label|true prediction rate|count|
+-----+--------------------+-----+
|0    |0.8                 |5    |
|1    |0.7333333333333333  |15   |
+-----+--------------------+-----+


-------------------------------------------
Batch: 7
-------------------------------------------
+-----+--------------------+-----+
|label|true prediction rate|count|
+-----+--------------------+-----+
|0    |0.7142857142857143  |7    |
|1    |0.75                |16   |
+-----+--------------------+-----+


-------------------------------------------
Batch: 8
-------------------------------------------
+-----+--------------------+-----+
|label|true prediction rate|count|
+-----+--------------------+-----+
|0    |0.7777777777777778  |9    |
|1    |0.7647058823529411  |17   |
+-----+--------------------+-----+
```

As you can see, the rates of true positive and true negative predictions are continuously updated as the data goes in. The true prediction rate is nothing out of this world because we are using a very small dataset and to make things worse, it had NULL values that have been discharged.

One of the main drawbacks of logistic regression is that it needs big datasets to be really able to get the insights of the data.

If you want to dig deeper into how to use Spark ML with Spark Structured Streaming, you can find a complete stream pipeline example following this link.

In the next section, we are going to analyze some of the expected future Spark Streaming features that are already here.

10.2 Spark 3.3.x

The new Spark 3.3.2 version was released on February 17, 2023, the time this book was written; therefore, some of the future improvements expected from Spark are already here[1].

For instance, one of the most recent Spark Streaming related improvements has been the addition of RocksDB state store provider, complementary to the default implementation based on the HDFS backend state store provider.

Although the incorporation of RocksDB as a state store provider is not new, it was included with Spark 3.2; RocksDB state store `WriteBatch` problems cleaning up native memory have been recently addressed.

Before Spark 3.2, the only built-in streaming state store implementation available was the HDFS backend state store provider (`HDFSBackedStateStore`). The HDFS state store implements two different stages. During the first phase, state data is stored in a memory map. The second phase includes saving that information to a fault-tolerance HDFS-compatible file system.

Remember from previous chapters that stream processing applications are very often stateful, and they retain information from previous events to be used to update the state of other future events.

When we have stateful operations such as streaming aggregations, streaming `dropDuplicates`, `stream-stream joins`, `mapGroupsWithState`, or `flatMapGroupsWithState` and, at the same time, we would like to maintain the state of a huge number of keys, this could cause processing latencies due to the problem of Java virtual machine (JVM) garbage collection, hence producing important delays in the micro-batch processing times.

To understand the reasons causing the previous state store problems, we have to know that the implementation of `HDFSBackedStateStore` causes the state information to be stored in the Spark executors' JVM memory. Thus, the accumulation of large numbers of state objects will saturate the memory originating garbage collection performance issues.

---

[1] `https://spark.apache.org/news/`

For situations like this, Spark recently incorporated RocksDB as another state storage provider, to permit storage of the state information in a RocksDB database.

Spark RocksDB State Store Database

Let's explore some of the new features RocksDB is bringing to the table and how you can use them to improve your Spark Streaming performance.

What Is RocksDB?

RocksDB is an embeddable persistent key-value store for fast storage based on three basic structures: `memtable`, `sstfile,` and `logfile`.

RocksDB includes the following main features:

- It uses a log structured database engine.

- It is optimized for storing small to medium size key-values, though keys and values are arbitrarily sized byte streams.

- It is optimized for fast, low latency storage such as flash drives and high-speed disk drives, for high read/write rate performance.

- It works on multicore processors.

Apart from Spark Streaming, RocksDB is also used as a state backend by other state-of-the-art streaming frameworks such Apache Flink or Kafka Streams which uses RocksDB to maintain local state on a computing node.

If you want to incorporate RocksDB to your Spark cluster, setting

```
spark.conf.set(
  "spark.sql.streaming.stateStore.providerClass",
"org.apache.spark.sql.execution.streaming.state.RocksDBStateStoreProvider")
```

enables RocksDBStateStoreProvider as the default Spark StateStoreProvider.

Apart from the previous basic configuration, Spark incorporates several options you can use to tune your RocksDB installation.

Table 10-2 includes a summary of the most common `RocksDBConf` configuration options for optimizing RocksDB.

Table 10-2. *RocksDB State Store Parameters*

Config name	Description	Default value
spark.sql.streaming.stateStore.rocksdb.compactOnCommit	When activated, RocksDB state changes will be compacted during the commit process	False
spark.sql.streaming.stateStore.rocksdb.blockSizeKB	Block size (in kB) that RocksDB sets on a BlockBasedTableConfig	4
spark.sql.streaming.stateStore.rocksdb.blockCacheSizeMB	The size (in MB) for a cache of blocks	8
spark.sql.streaming.stateStore.rocksdb.lockAcquireTimeoutMs	Milliseconds to wait for acquiring lock in the load operation for a RocksDB instance	60000
spark.sql.streaming.stateStore.rocksdb.resetStatsOnLoad	Whether all ticker and histogram stats for RocksDB should be reset on load	True
spark.sql.streaming.stateStore.rocksdb.trackTotalNumberOfRows	Whether to track the total number of rows Performance issues It adds additional lookups on write operations, however enableing it could represent a possible RocksDB Performance Degradation	True

BlockBasedTableConfig

BlockBasedTable is RocksDB's default SST file format. It includes the configuration for plain tables in `sst` format. RocksDB creates a `BlockBasedTableConfig` when created.

RocksDB Possible Performance Degradation

With this option enabled, it adds extra attempts to retrieve data on write operations to track the changes of the total number of rows, bringing an overhead on massive write workloads. It is used when we want RocksDB to upload a version of a pair key-value, update the value, and after that remove the key. Thus, be advised turning it on can jeopardize the system performance.

Wrapping up, RocksDB is able to achieve very high performance. RocksDB includes a flexible and tunable architecture with many settings that can be tweaked to adapt it to different production environments and hardware available, including in-memory storage, flash memory, commodity hard disks, HDFS file systems, etc.

RocksDB supports advanced database operations such as merging, compaction filters, and SNAPSHOT isolation level. On the other hand, RocksDB does not support some database features such as joins, query compilation, or stored procedures.

10.3 The Project Lightspeed

On June 28, 2022, Databricks announced the Project Lightspeed. The Project Lightspeed is the next-gen Spark Streaming engine.

Spark Structured Streaming has been widely adopted by the industry and community, and as more and more nowadays applications require processing streaming data, the requirements for streaming engines have changed as well.

The Project Lightspeed will focus on delivering higher throughput and lower latency and reduce data processing cost. Project Lightspeed will also support the expansion of the ecosystem of connectors, enhance new streaming functionalities, and simplify the application deployment, monitoring, and troubleshooting.

Project Lightspeed will roll out incrementally, backing the improvement of the following Spark Structured Streaming fields:

Predictable Low Latency

In this field, the new Apache Spark Structured Streaming is promising the increase of workload performance as much as twice in comparison with current capabilities. Some of the initiatives that are currently taking place in this area to support the consecution of this objective are as follows:

- Offset management: Practical experience shows that offset management operations consume 30% to 50% of the total pipeline processing time. It is expected to reduce processing latency by making these operations asynchronous and of configurable pace.

- Asynchronous checkpointing: It is expected up to 25% of improvement in efficiency in this domain by overlapping the execution with the writing of the checkpoints of two adjacent groups of records. Currently checkpoints are written only after processing each group of records.

- State checkpointing frequency: New Spark Structured Streaming engine is expected to incorporate the parametrization of the number of checkpoints, that is, writing checkpoints only after processing a certain number group of records in order to reduce latency.

Enhanced Functionality for Processing Data/Events

Project Lightspeed is going to enlarge Spark Structured Streaming functionalities in the following fields:

- Multiple stateful operators: The new Spark Structured Streaming is expected to support multiple state operators in order to satisfy multiple use cases such as the following:

  o Chained time window aggregation (e.g., chain of different types of window aggregations)

  o Chained stream-stream outer equality joins.

  o Stream-stream time interval join plus time window aggregations

- Advanced windowing: The new Spark Structured Streaming engine is expected to provide an intuitive API to support the following:

  ○ Arbitrary groups of window elements

  ○ Ability to define when to execute the processing logic

  ○ Capacity to remove window elements before or after the processing logic is triggered

- State management: Lightspeed will incorporate a dynamic state schema adapting to the changes in the processing logic and the capacity to externally query intermediate information ("state").

- Asynchronous I/O: Lightspeed is also expected to introduce a new API to asynchronously manage connections to external data sources or systems. This new functionality can be very helpful in streaming ETL jobs that collect live data from heterogeneous sources and/or writing into multiple sinks.

- Python API parity: Lightspeed will provide a new Python API incorporating stateful processing capabilities and built-in integrations with popular Python packages like Pandas to facilitate its utilization by Python developers.

New Ecosystem of Connectors

Connectors certainly make Spark users' life easier. In this area, the Project Lightspeed is also expected to supply the following:

- New connectors: New native connectors will be added to Spark Structured Streaming. For example

  ○ Google Pub/Sub, which is an asynchronous and scalable messaging service acting as an interface between services producing messages and services processing those messages

  ○ Amazon DynamoDB, which is a NoSQL database service

- Connector enhancement: New functionalities are also expected in this area, such as including AWS IAM auth support in the Apache Kafka connector or enhancement of the Amazon Kinesis connector.

Improve Operations and Troubleshooting

Processing unbound data requires the system treating that information to be up and running 24/7. Therefore, constant monitoring and managing of those types of systems while keeping operating costs under control is incredibly relevant to business. Thus, as part of Project Lightspeed, Spark is anticipated to incorporate two new set of features:

- Observability: Structured streaming pipelines will incorporate:

  - Additional metrics for troubleshooting streaming performance.

  - The mechanism for collection metrics will be unified.

  - The capacity to export metrics data to different systems and formats will be enhanced.

  - Visualization tools will be improved.

- Debuggability: As in the previous point, structured streaming pipelines will also integrate capabilities to visualize the following:

  - How pipeline operators are grouped and mapped into tasks.

  - The tasks running on the executors.

  - Executors' logs and metrics drill down analysis capacity.

10.4 Summary

In this chapter, we discussed the capacities of Spark Structured Streaming when coupled with Spark ML to perform real-time predictions. This is one of the more relevant features the Apache Spark community is expected to improve as more business applications require in-motion data analytics to trigger prompt reaction. After that, we discussed the advantages of the new RocksDB State Store to finalize with one of the most expected Spark Streaming turning points, the Project Lightspeed, which will drive Spark Structured Streaming into the real-time era.

Bibliography

- Akidau, T., Chernyak, S., & Lax, R. (2018). *Streaming Systems: The What, Where, When, and How of Large-Scale Data Processing* (first edition). O'Reilly Media.

- Chambers, B., & Zaharia, M. (2018). *Spark: The Definitive Guide: Big Data Processing Made Simple* (first edition). O'Reilly Media.

- Chellappan, S., & Ganesan, D. (2018). *Practical Apache Spark: Using the Scala API* (first edition). Apress.

- Damji, J., Wenig, B., Das, T., & Lee, D. (2020). *Learning Spark: Lightning-Fast Data Analytics* (second edition). O'Reilly Media.

- Elahi, I. (2019). *Scala Programming for Big Data Analytics: Get Started with Big Data Analytics Using Apache Spark* (first edition). Apress.

- Haines, S. (2022). *Modern Data Engineering with Apache Spark: A Hands-On Guide for Building Mission-Critical Streaming Applications*. Apress.

- *Introducing Native Support for Session Windows in Spark Structured Streaming*. (October 12, 2021). Databricks. www.databricks.com/blog/2021/10/12/native-support-of-session-window-in-spark-structured-streaming.html

- Kakarla, R., Krishnan, S., & Alla, S. (2020). *Applied Data Science Using PySpark: Learn the End-to-End Predictive Model-Building Cycle* (first edition). Apress.

- Karau, H., & Warren, R. (2017). *High Performance Spark: Best Practices for Scaling and Optimizing Apache Spark* (first edition). O'Reilly Media.

© Alfonso Antolínez García 2023
A. Antolínez García, *Hands-on Guide to Apache Spark 3*, https://doi.org/10.1007/978-1-4842-9380-5

- Kukreja, M., & Zburivsky, D. (2021). *Data Engineering with Apache Spark, Delta Lake, and Lakehouse: Create Scalable Pipelines That Ingest, Curate, and Aggregate Complex Data in a Timely and Secure Way* (first edition). Packt Publishing.

- Lee, D., & Drabas, T. (2018). *PySpark Cookbook: Over 60 Recipes for Implementing Big Data Processing and Analytics Using Apache Spark and Python* (first edition). Packt Publishing.

- Luu, H. (2021). *Beginning Apache Spark 3: With DataFrame, Spark SQL, Structured Streaming, and Spark Machine Learning Library* (second edition). Apress.

- Maas, G., & Garillot, F. (2019). *Stream Processing with Apache Spark: Mastering Structured Streaming and Spark Streaming* (first edition). O'Reilly Media.

- *MLlib: Main Guide—Spark 3.3.2 Documentation.* (n.d.). Retrieved April 5, 2023, from `https://spark.apache.org/docs/latest/ml-guide.html`

- Nabi, Z. (2016). *Pro Spark Streaming: The Zen of Real-Time Analytics Using Apache Spark* (first edition). Apress.

- Nudurupati, S. (2021). *Essential PySpark for Scalable Data Analytics: A Beginner's Guide to Harnessing the Power and Ease of PySpark 3* (first edition). Packt Publishing.

- *Overview—Spark 3.3.2 Documentation.* (n.d.). Retrieved April 5, 2023, from `https://spark.apache.org/docs/latest/`

- Perrin, J.-G. (2020). *Spark in Action: Covers Apache Spark 3 with Examples in Java, Python, and Scala* (second edition). Manning.

- *Project Lightspeed: Faster and Simpler Stream Processing with Apache Spark.* (June 28, 2022). Databricks. `www.databricks.com/blog/2022/06/28/project-lightspeed-faster-and-simpler-stream-processing-with-apache-spark.html`

- Psaltis, A. (2017). *Streaming Data: Understanding the real-time pipeline* (first edition). Manning.

- *RDD Programming Guide—Spark 3.3.2 Documentation.* (n.d.). Retrieved April 5, 2023, from `https://spark.apache.org/docs/latest/rdd-programming-guide.html`

- Ryza, S., Laserson, U., Owen, S., & Wills, J. (2017). *Advanced Analytics with Spark: Patterns for Learning from Data at Scale* (second edition). O'Reilly Media.

- *Spark SQL and DataFrames—Spark 3.3.2 Documentation.* (n.d.). Retrieved April 5, 2023, from `https://spark.apache.org/docs/latest/sql-programming-guide.html`

- *Spark Streaming—Spark 3.3.2 Documentation.* (n.d.). Retrieved April 5, 2023, from `https://spark.apache.org/docs/latest/streaming-programming-guide.html`

- *Structured Streaming Programming Guide—Spark 3.3.2 Documentation.* (n.d.). Retrieved April 5, 2023, from `https://spark.apache.org/docs/latest/structured-streaming-programming-guide.html`

- Tandon, A., Ryza, S., Laserson, U., Owen, S., & Wills, J. (2022). *Advanced Analytics with PySpark* (first edition). O'Reilly Media.

- Wampler, D. (2021). *Programming Scala: Scalability = Functional Programming + Objects* (third edition). O'Reilly Media.

Index

A

Accumulators, 104, 105
Adaptive Query Execution (AQE)
    cost-based optimization framework,
        196, 197
    features, 197
    join strategies, runtime
        replanning, 200
    partition number, data dependent,
        198, 199
    spark.sql.adaptive.enabled
        configuration, 201
    SQL plan, 198
    unevenly distributed data joins,
        200, 201
Apache Parquet, 123
Apache Spark
    APIs, 12
    batch *vs.* streaming data, 17, 18, 20, 21
    dataframe/datasets, 12, 13
    definition, 3
    execution model, 11
    fault tolerance, 6, 7
    GraphX, 16
    scalable, 5
    streaming, 14, 15
    unified API, 7
Apache Spark Structured Streaming,
    249, 333
Application programming interface
    (API), 3, 67
.appName() method, 75

B

Batch data processing, 19, 207
Broadcast variables, 102–104, 106

C

Cache(), 187, 189, 190
Checkpointing streaming
    example, 284
    failures, 287
    features, 284, 285
    state data representation, 286
    stateful, 283
    state storage, 286
collect() method, 73
collect() method, 76
countByValue() method, 99
count() function, 219
createDataFrame() method, 116, 117
Customer Relationship Management
    (CRM), 7
customWriterToConsole() function, 324

Artificial intelligence (AI), 367
Association for Computing
    Machinery (ACM), 7

D

Data analytics, 18, 204, 331
DataFrame API, 109
    Apache Parquet, 123, 124

397

A. Antolínez García, *Hands-on Guide to Apache Spark 3*, https://doi.org/10.1007/978-1-4842-9380-5

Printed in the United States
by Baker & Taylor Publisher Services